Franciscans and the Elixir of Life

THE MIDDLE AGES SERIES

Ruth Mazo Karras, Series Editor
Edward Peters, Founding Editor

A complete list of books in the series
is available from the publisher.

FRANCISCANS AND THE ELIXIR OF LIFE

Religion and Science in the Later Middle Ages

ZACHARY A. MATUS

PENN

UNIVERSITY OF PENNSYLVANIA PRESS

PHILADELPHIA

Published by
University of Pennsylvania Press
Philadelphia, Pennsylvania 19104-4112
www.upenn.edu/pennpress

Printed in the United States of America on acid-free paper
1 3 5 7 9 10 8 6 4 2

Library of Congress Cataloging-in-Publication Data
Matus, Zachary A., author.
Franciscans and the elixir of life : religion and science in the later Middle Ages / Zachary A. Matus—1st ed.
Includes bibliographical references and index
Philadelphia : University of Pennsylvania Press, [2017]
pages cm. (The Middle Ages series)
ISBN 9780812249217 (hardcover : alk. paper)
1. Franciscans—History—To 1500. 2. Alchemy—Religious aspects—Christianity—History—To 1500. 3. Religion and science—Europe—History—To 1500. 4. Elixir of life. I. Title. II Series: The Middle Ages series
BR115.A57 M38 2017 2016053771

For Suzanne

CONTENTS

Introduction

Producing the elixir of life was one of two major aims of medieval alchemists. Metallurgical alchemy, the transmutation of base metals, usually into gold or silver, was the other. Often discussed as a pseudoscience, alchemy in fact played a significant part in the genealogy of modern chemistry. It dealt, above all, with matter—its manipulation, improvement, and general properties. Sometimes limited to techniques that would be known to dyers, metal workers, and other artisans, in its most elaborated form alchemy was a *scientia* that explained the composition of the physical universe. Alchemy was tied quite closely to other disciplines of natural philosophy, including physics, astrology, and medicine. Yet in spite of its putative ability to explain the composition of material things, alchemy, unlike its sister disciplines, never gained a lasting foothold in the schools.

Perhaps because of this development, alchemy was not standardized. There was no single definition, nor a general curriculum. There were influential works, but as a practice outside or at the fringe of the university, medieval alchemy was idiosyncratic. Unlike, for instance, the study of theology or academic medicine, where students were expected to annotate specific texts with their master's commentary, the decision to write about or practice alchemy was very much an expression of individual preference and circumstance. Therefore, it was not just detractors who argued with adherents over definitions of alchemy and its place within the fields of medieval *scientiae* and, more broadly, its proper role in Christendom. Adherents as well seldom agreed with one another on these questions. This is not without advantage to the historian, however. Alchemy's marginality refracts, rather than reflects, normative intellectual life. It provides us a better perspective through which to understand the intellectual culture of the era, precisely because alchemical literature resists essentialization and generalization. This disunity of the literature was apparent enough that by the later Middle Ages, alchemical schools such as the Pseudo-Lullian recognized the messy reality of prior generations and sought to solve the problem through interpolations and elisions in the manuscript tradition.

Discord, however, required some common ground on which alchemical ideas could be debated. Over the course of the twelfth and thirteenth centuries, the practitioners and theoreticians of alchemy organized their discipline on then contemporary and widely accepted principles of natural philosophy.[1] Unlike philosophy, however, the theorization of alchemy often included a type of empirical practice. Medieval adherents of alchemy relied on observations and tests called *experimenta* (or sometimes *documenta*).[2] Robert Bartlett suggests that we might best translate terms like *experimenta* and *experientia* to mean observations, rather than experiments, in order to avoid confusion with the modern terminology. Still, what is important is that alchemists took into consideration the results of their practice, or the practice of others, rather than relying exclusively on *argumentum* (reasoning).[3] This is not the same as saying that the sort of philosophical reasoning common to medieval philosophy was a subsidiary concern to physical trials. It is better to say that alchemical discourse, like natural philosophy, was founded on both reasoned argument and established opinion, but could—and did—account for alchemical praxis to inform its philosophical conclusions. Therefore, like chirurgery and empirical medicine, alchemy occupied space between the "liberal and manual arts" and consequently was held in less esteem than many of its sister disciplines.[4]

Latin commentators of the era—be they translators, practitioners, or skeptics—often referred to alchemy as a *novitas* (a novelty), a term that could connote disdain, but also signaled to the intellectual community the opening of a new scholarly question and endeavor.[5] While it is true that some of the techniques and processes that made up the alchemical craft were known in the West well before the twelfth century, alchemy as a distinct branch of knowledge was no longer differentiated as such in the Latin West after the upheavals of late antiquity. It reemerged as a specific discipline over the course of the twelfth and early thirteenth centuries. What had remained in the early Middle Ages and into the beginning of the High Middle Ages was a collection of lore, recipes, techniques, and strands of classical medical theory—none of which operated under the formal rubric of alchemy. The Arab inheritors of antique alchemy fused the ancient Neoplatonic and Hermetic alchemical practices to Aristotelian natural philosophy, allowing alchemy to emerge as a coherent discourse, even if its adherents and detractors did not agree as to its precise potential or the justification for it.[6] In the twelfth century the massive translation project that brought Aristotelian philosophy, the *scientia* of heavenly bodies and their motion (astronomy or astrology), and classical medicine into Western hands also brought with it alchemical texts.

Yet, what occurred was not simply translation. The magnitude of the importation also transformed alchemy. Western academics developed new theories, wrote their own texts, and brought their own cultural assumptions and prejudices to bear on the development of the *scientia* of alchemy. Aristotelian notions such as soul, essence, form, and material littered the landscape of the new alchemy, but did so weighted with meaning unintended by the Stagirite. Aristotle used a host of terms such as these as *concepts* meant to describe the world. In the alchemical texts of both Arabic and Latin provenance, however, these conceptual terms were used to discuss actual substances that could be manipulated in the laboratory. Essence was not an idea; it was a substance that could be identified, distilled, and manipulated.[7] It is also important to bear in mind that, just as Aristotle never fully developed his chemical geology, neither did he discuss the possibility of transmutation. Therefore, the medieval alchemist, Islamic or Latin, often innovated or, at the very least, extrapolated to fill lacunae in the Aristotelian corpus.

Medicinal alchemy was largely a Western phenomenon. Terminology and essential theory were borrowed from translated treatises from the Islamic world, but the desire to use alchemy to make medicine was not a traditional theme of Islamic alchemists (though Islamic physicians certainly were interested in the possibility of making compound drugs). The elixir of life was the alchemical medicine par excellence, though its composition and precise properties tended to vary from writer to writer. Though the differences among theories of and instructions for making the elixir often prove revealing, it is possible to recognize some more or less general qualities of the substance. It was a universal medicine or cure-all, as well as a means by which one could significantly extend human life. The elixir did not promise immortality, however. Immortality was a quality reserved for God and the resurrected dead after the Last Judgment, though some alchemists blurred this distinction.

Given the pervasive, though hardly dominant clerical discourse that care for the body often came at the cost of care for the soul, it might seem surprising that a handful of Franciscans might give such a vaunted place to the elixir. The elixir, however, did not instill physical benefits only. It also could endow one with gifts that might be better described as spiritual, emotional, or intellectual. It could ease spiritual suffering. It cleared the mind and instilled confidence, bravery, and even intelligence. Just as metallurgical alchemists transmuted and ennobled a base metal into gold by removing its negative properties and instilling positive ones, so too did alchemists hope to purify, transmute, and ennoble the human body. The elixir's ability to rewrite the composition of the human body transcended the power of most medicines. Nevertheless, the elixir was closely related to the development of compound drugs. The study and production of compound medicines, some of which required some artisanal skill to create, were part of the formal medical curriculum at the universities. Yet caution pervaded such discussions. At the University of Montpellier, for example, the foremost medical school of the later Middle Ages, masters were aware of the potential harm of new medicaments. Yet they allowed for the combination of drugs to also exhibit new properties that could not be explained by the combination of their basic ingredients.[8] The compound drug called the *theriac*, for instance, rivaled the elixir in its potency according to some of its proponents, and, as I will discuss in Chapter 2, there can be some overlap between the theriac and the elixir. The general aim of even the most powerful compound drugs was to heal and repair certain conditions or illnesses. The elixir, however, transformed.

My study focuses principally on three Franciscans interested in the elixir. For two of them, Roger Bacon and John of Rupescissa, the elixir had the potential to reshape not only bodies, but the world. For the third "alchemist," Franciscan cardinal and plagiarist Vitalis of Furno, the elixir was a powerful medicine and potentially transformative, but ultimately mundane. Spanning the mid-thirteenth to the mid-fourteenth century, the history of all these men and their alchemical works was tangled up with the turbulent history of the Franciscan Order. From the early flirtations with the apocalyptic works of Joachim of Fiore and the controversy over poverty, through the burgeoning of the Spiritual problem, to the immediate aftermath of Pope John XXII's attacks on the Spirituals and on the core notion of Franciscan poverty itself, the Order was rent into pieces by conflicts both within and without. These troubles left their mark on each of the principal figures of this study.

I have tried, therefore, to balance the larger alchemical discourses with the individual features of the texts I have engaged in this study. Roger Bacon is among those historical figures whose legend outshines historical data.[9] Many of the once salient details of Bacon's life—persecution, arcane masteries, quintessential Englishness—emerge from rather unreliable sources at least a century after his death. We do know he was born sometime around 1214 or

1220, and died around 1292. Bacon finished his studies in Oxford probably by the mid-1230s, and may himself have been a master at Paris in the 1240s or earlier, though he was apparently back at Oxford shortly thereafter.[10] Having already pursued an active academic career, he entered the Order of Friars Minor around 1256, just prior to Bonaventure's elevation to the post of Minister General. Bacon's admission came in the wake of the so-called scandal of the Eternal Gospel, when, in 1254, the young Franciscan Gerardo of Borgo San Doninno penned a text called the *Introductorius*, in which he claimed that Joachim of Fiore's prophetic-exegetical works superseded scripture. This misstep provided the rivals and enemies of the Franciscans an opening through which to attack the Order. It is therefore difficult to countenance claims that Bacon took the Franciscan habit strictly as a means of advancement.

Bacon met Cardinal Guy Foulques, later Pope Clement IV, in 1263 or 1264. It was in the course of this conversation that Bacon first mentioned to the future pope the topics he would later include in his *Opus maius* (*The Greater Work*) and in the recapitulations and emendations of the *Opus minus* (*The Lesser Work*) and *Opus tertium* (*The Third Work*). In 1265, Guy became pope and the next year ordered Bacon to send him a treatise covering the issues they had discussed, as well as his insights on philosophy, theology, the secular-mendicant controversy at Paris, and the role of Aristotle. Bacon complied by hastily writing the *Opera* and neglected to submit it to his Franciscan superiors beforehand. A late fourteenth-century source claims that Bacon was condemned between 1277 and 1279. The cause of the condemnation is not known, except that it was due to various "novelties." Bacon's work offers a number of possible candidates for such a distinction, so it is certainly possible. In any case, around 1278 Bacon had departed Paris for Oxford, and the events of the years until the end of his life are unknown apart from the completion of a few works.

Vitalis of Furno's reputation as an alchemist is almost nonexistent. He may not have even authored the encyclopedic treatise where his discussion of the elixir is found, though the book's Franciscan provenance seems secure. Vitalis was better known as a careerist, whose academic interests seem to have existed only to serve his desire to climb the ecclesiastical hierarchy. He was born in Gascony sometime before 1260, but little is known of Vitalis's life prior to his entrance into the university. We know he began his studies in theology at Paris around 1285, where he read the *Sentences* under Jacques de Quesnoy.[11] We do not know when he entered the Order, but he returned to

Montpellier as a lector at the Franciscan *studium* in 1291, and a few years later returned to Paris. In the closing years of the century he was lector at Toulouse, though Stephen Marrone points to 1300 as the effective end of his academic career. Still, it was far from the end of his intellectual life. In 1307 he was named Provincial in Aquitaine. In 1312 Clement V named him cardinal priest of St. Martin in Montibus, and in 1321 John XXII elevated him to cardinal bishop of Albano. He died in 1327.[12] Vitalis was a staunch supporter of the conventual movement and was a strong ally of the pope almost to the end of his life. Like many of his confreres, Vitalis objected to John XXII's attacks on the Franciscan understanding of the absolute poverty of Christ. He rather spectacularly fell out of favor when the pope viciously harangued him in consistory.

John of Rupescissa was born around 1310.[13] We do not know much about his early life, except that he was born into the lesser nobility in Aurillac and obviously procured enough education for himself to enroll at the University of Toulouse in 1327. Later in life he remembered his university years as a tumultuous time, where he was caught up in worldly things. John remained at the university for five years before joining the Order. In the same year he took the Franciscan habit he claims to have had his first vision of Antichrist. John initially kept his visions to himself, only revealing them some years later in a Franciscan chapter meeting in 1335 or 1336, along with his prediction of hostilities between England and France, which led eventually to his fame as a prophet of the Hundred Years War.

John's initial reluctance to share his visions appears to have been justified, for his Franciscan superiors did not share his conclusion that God desired him to speak of what he had seen. By 1344, these superiors, led by Guillaume Farinier, provincial of Aquitaine, had heard enough. They confined John to a small cell in the Franciscan convent of Figeac. John remembers the time in Figeac as one of great suffering. He was placed on a rack for treatment of a broken leg, and left bound in irons to rot in his own filth for more than three months, with maggots crawling over his infected wounds. John survived, strengthened in his own belief that God had granted to him a special enlightenment. The bodily trials of Figeac subsequently gave way to trials of a legal nature. John was shuttled between various Franciscan convents and was tried for heresy for the first time in 1346 by a local Dominican inquisitor. The conclusion of the inquisition apparently did not suit Farinier, who rearrested John and sent him to be imprisoned in Toulouse and later in Rieux.

John's sufferings continued. He contracted the plague while in prison, prompting more visions. But again, he survived. He eventually managed to convince a sympathetic captor to take him to the papal court at Avignon instead of to the next in the lengthening series of Franciscan convents that served as his jails. At Avignon, John was treated better, and as part of his examination by the cardinal inquisitor, he was asked to compose a lengthy description of his visions of the future, rather than merely to submit whatever previous writings he had with him. After much debate, and after a trial in which John roundly condemned the excesses of the clergy, he finally was declared *fantasticus,* but not *hereticus*. Upon this finding, he was remanded again to prison, where he would stay until he was allowed to enter the local Franciscan convent just prior to his death. Though John railed against his jailing, the papal prison in Avignon proved an excellent writer's workshop. It was there that he composed all his surviving works. Indeed, John became a fecund author, with more than twenty books written before his death, and he was frequently sought out for prophetic advice.

Each of these friars contributed to alchemical discourse and the elixir tradition in his own way, yet they also represent a particular period in the history of alchemy. I consider only these works from the thirteenth and fourteenth centuries, not simply as a means to limit the scope of the project, but also because the turn of the fifteenth century heralded a shift in alchemical writing and praxis. It was around this time when, as Pamela Smith has noted, there was a surge in the production of technological books written in the vernacular—a change in the production of alchemical treatises that intensified with the invention of printing.[14] In the wake of these developments, alchemy became a much more widespread practice both in the marketplace and at home, changing the character of the discipline.

* * *

It is worth asking whether Bacon, Vitalis, and John were tied together by something more than their membership in the Friars Minor and their interest in the elixir. Was there a Franciscan alchemy? If by Franciscan alchemy we mean something along the lines of a school, I think not. On the other hand, lived Franciscanism—insofar as each author understood or experienced it—played a significant role in the alchemical works of each. It is the nexus between Franciscanism and alchemy that is the subject of this book. It is through the particularities of these relationships that broader connections and ruptures between medieval religion and science come into view.

I have conceptualized this study, therefore, as a history *around* science. At its most basic, it describes and situates the pursuit of the elixir within the intellectual, religious, and political culture of the friars who wrote about it.[15] Taking this method further, I present three related, but distinct arguments. First, writing about alchemy—no matter how few Franciscans did so—did not require a radical intellectual break with the rest of Franciscan intellectual culture. This is especially true of writings on the elixir, but can be applied more broadly to alchemy as a whole. Second, for the friars of this study alchemy's materiality and attention to the physical world was what conspicuously linked it to religion. Alchemy's focus on the corporeal was neither an impediment nor superfluous to its connection to religion, nor was it a degradation of an erstwhile purely spiritual experience. Instead, the development of these Franciscan alchemies emerged in a period when Christians across Europe were focusing on and seeking for material evidence of their faith and material means of practicing their devotion. The doctrines of creation, incarnation, and resurrection dealt inherently with the material world, but more to the point Christian *practice* was thoroughly intertwined with matter. The Eucharist, baptism, devotional art, relics and reliquaries, and even the smells and sounds of the Mass speak to the importance of the corporeal.[16] Third, the religious and even the liturgical world of the Franciscans left an impact on their alchemical works.

The intellectual genealogy of this latter argument owes much to the work of Leah DeVun and Agostino Paravicini-Bagliani, both of whom have explored this topic as it relates to John of Rupescissa and Roger Bacon, and each has demonstrated the importance of reading alchemical texts in light of an author's larger oeuvre and other circulating discourses.[17] Adopting this method myself, I also want to add what may seem a counterintuitive argument—that not only did religious *thought* influence alchemical writings, but also religious *practice.* By practice I mean principally Franciscan ritual life. Ritual life was frequently formalized into liturgical structures, but these structures did not necessarily limit improvisation and imagination. Too often ritual is understood in terms of things that are "ritualistic," overemphasizing formulas or ossified repetitions. Here, I mean ritual as a mode of being in the world. Following the work of J. Z. Smith and, more recently, of Adam Seligman, Robert Weller, Michael Puett, and Bennett Simon, I focus on ritual as a means of approaching the world subjunctively, that is, as it could or should be. Like the ritual itself, subjunctivity is ephemeral and fleeting, but it privileges the experience of meaning and connection. Franciscan ritual life therefore had implications for how one "thought" alchemy, for it could explain both the extraordinary potential of alchemical science and the inevitable clash between potentiality and the empirical inability of the elixir to achieve its reputed effects. Alchemy as a subjunctive science expresses conditional truths.

The fact that religious and scientific thought are deeply imbricated with one another in discussions of the elixir points to something deeper than a mere overlay of religious language onto alchemical theory. While it is possible to disentangle the alchemical praxis (the production of a chemical compound) from religious aspirations, the character of the substances created—the nature of the elixir of life—is deeply influenced by theological and ritual assumptions that are too tightly entangled in the science to be unwound. Moreover, there is a real danger of fundamentally misunderstanding the historical actors and their ideas if we attempt to remove the scientific marrow and examine it separately from its context.[18]

Medieval distinctions in genre have little to do with separating "science" from "religion," in any case. Both terms—religion and science—are in fact anachronisms when applied to the Middle Ages. The basic meaning of *scientia* was knowledge, but the actors in this study better understood *scientia* a field of knowledge that was systematically organized and intelligible via the application of reason. Hence, theology and ethics were *scientiae* just as was optics or (another difficult term to translate) *astronomia*. Likewise *religio* meant something like an obligation to worship. The term "religious" in the Middle Ages tended to refer not to people of a spiritual frame of mind, but to men and women who lived under a religious rule—a designation that included a wide range of categories to be sure, but also one that limited the religious to a place within the hierarchy of the Christian Church. Much as we read "science" back into the Middle Ages, often by substituting the term for what medievals called natural philosophy, likewise we read "religion" back into the Middle Ages for a wide variety of texts and behaviors. My goal is not to muddy the waters here by claiming that modern terminology is of no value, but simply to recognize that modern terms generate modern assumptions about division of knowledge that do not necessarily hold true when applied to medieval Europe.

Science is, to adapt a theory from Dipesh Chakrabarty, a colonizing and totalizing discipline.[19] Historians now recognize that the development of science, in the modern sense, was hardly a necessary series of events and, moreover, that it developed in fits and starts quite different from the clear narrative of progress proposed by scholars less than a century ago. Yet, for all of our recognition of proto-scientific practices that include nonscientific qualities, the logic of science nevertheless dictates that we make this very distinction, that the practice of alchemy can include both scientific and nonscientific aspects. To put it another way, the existence of science today presupposes a past where its antecedents existed. This may seem something of a tautology, but the critical issue at stake here is that such an observation implies a kind of division in the practice of alchemy that would not have been recognizable to its adherents. Science per se never existed—nor does it now—as anything more than a discourse. Discourses are powerful things, to be sure, but they are always contested, and, just as importantly, they emerge in the context of specific individuals, each of whom is positioned within a correspondingly specific cultural and historical matrix and freighted by individual experiences. By retrojecting the discourse of science into the past we draw distinctions that are useful to understanding the genealogy of the present, but tend to obfuscate in trying to recreate historical thought worlds.

Something similar could be said about casting modern definitions of religion into history. During the Enlightenment and well into the twentieth century, a number of scholars understood religion to have developed on a teleological path (usually toward Protestant Christianity or atheism). While this view is rather muted today, religion is even more likely than science to be essentialized. The movement to study lived religion has done much to capture the richness of religious practice, but Christianity is still used as something of a monolithic term. It is not a given that Christians of the thirteenth or the fourteenth century are intrinsically recognizable to Christians of the twenty-first, or that outside observers would classify them as a group. The survival of a tradition is not enough on its own to presume its intelligibility. Likewise, we need not expect uniformity when it comes to religious practice or belief. The obsession of the clerical class to regulate religious behavior in the later Middle Ages is proof enough that the ideal of uniformity remained an ideal only. Medieval alchemical texts also show, rather less ostentatiously, their own divergence on questions of religion. Even after what William Newman has termed the "religious turn"—an efflorescence of religious language in alchemical texts—in the early fourteenth century, great numbers of what we might call naturalist texts continued to be produced.[20] These texts might make no mention of God at all, or, at most, frame the alchemical pursuit within a universe created and governed by God, who acted more or less as simply a guarantor and source of natural laws.

While this is sufficient, I think, to qualify as a religious idea, it is qualitatively different from characterizing alchemy as divinely revealed knowledge.

One of the problems posed by religious language is that it has led to some confusion as to the principal aim of the alchemist. The history of religion has a particularly troubled historiography on this score, rather in contrast with the work of historians of science in recent decades, who have demonstrated quite convincingly that alchemy focused on the manipulation of matter.[21] In the 2004 *Encyclopedia of Religion* under the heading "alchemy," the reader will find the statement by Mircea Eliade for the 1987 edition that "alchemy was not scientific, but spiritual."[22] Eliade was far from the first to make this claim, preceded in the twentieth century by psychoanalyst Carl Jung, who posited that the coded language of alchemy symbolized an untranslatable psychic or mystical journey. Eliade, in his work *The Forge and the Crucible*, expressed the idea that alchemy was a spiritual process aimed not at the transformation of matter, but at that of the soul. Eliade concentrated on the experience of alchemical creation, equating it with a mystical, even gnostic, religiosity that was consciously ahistorical and impenetrable.[23] In addition to mischaracterizing the physical manipulation of matter as superfluous—transmuting physical substances was always at the heart of alchemy, even if the compound itself had soteriological implications—Eliade also narrowed the sense of religion to a kind of irrational mystical experience. Irrationality is not in and of itself the problem. Historians have had some success when it comes to dealing with the putatively irrational. Emotions, sexuality, mysticism, and madness have all been brought into the domain of historical study, even if debates about conclusions and methodology remain. Instead, when alchemy is characterized as a form of esoteric mysticism that denies the importance of or seeks to transcend the material world, the picture of how religion and religious persons intersected with alchemy becomes drastically skewed. This essentialized view of religion omits theology, ritual, scripture, and, for good measure, historical context. If religious experience is limited to an ephemeral, intangible, incommunicable experience with "real" presence, there remains no space for the materiality of alchemy to influence religious thought or behavior and vice versa. Therefore, this study works with religion as a cultural and historical category rather than an exogenous definition of what "real" religion is. In so doing, the connections between alchemy and religion come into sharper focus.

Materiality, of course, is not the only connection to be made between alchemy and religion. By the turn of the fourteenth century, many alchemical writers specifically framed their works with Christian ideas about the nature of God and the universe to make it easier, both morally and metaphorically, to grasp its essential principles. A good example of this phenomenon comes from the fourteenth-century *Codicil*, which compares the process of alchemical transmutation to the role of Christ in salvation: "Just as Christ assumed human form to liberate mankind from the transgression of Adam and its imprisonment by sin, so is it in our art. What is foully stained by one thing is cleansed by its opposite from that wicked taint . . . so approaching the limit of perfection, it becomes more perfect."[24] Barbara Obrist has argued that allegory, metaphor, and analogy served a rhetorical function. Rather than being a means of concealing knowledge, the florid imagery, allegory, metaphor, and specifically Christian themes helped render alchemical precepts understandable to a new audience. For a number of authors of this period, however, alchemy was not simply described in Christian nomenclature, but was fused to Christian conceptions of the universe.[25] The need to explicate alchemy, then, is not sufficient in and of itself to explain the rise in religious language and ideas occurring in alchemical texts. We must look also at the discourses that were co-opted in the course of the flourishing of religious alchemical texts.

My method, then, is to read Franciscan elixir texts within the context of related discourses. This allows me to examine the relationship of religion and science within the conceptual frameworks native to the period, such as alchemy, natural philosophy, and apocalypticism. In so doing, I seek to avoid essentializing science and religion without losing either as a constructive, critical category. This particularist reading operates on two basic assumptions, each of which will be borne out throughout the discussion. The first is that discourses are porous. Genre and other discursive boundaries might restrict the content or presentation of a work, but they do not forbid influence (conscious or otherwise) from one field to another. This seems intuitively true for any individual author, but we also have evidence, for instance, from the thirteenth-century Franciscan chronicler Salimbene de Adam that friars frequently were exposed to new or controversial opinions through conversation and disputation.[26] It is important to attend, then, to the range of informal and formal discourses across the network of the Franciscan Order. Second, a genealogical model is more productive for understanding connections between religion and alchemy than is a developmental or a synthetic model. These Franciscans built their alchemies from many sources—Aristotelian commentaries, theological claims, and scriptural writings—but not always from the same sources, and they did not draw identical conclusions. Important similarities and themes emerge among various writers that are respective of their context, but the works fit comfortably neither in a narrative of progress nor as an expression of a singular idea.

Historical context is the subject of Chapter 1. In this chapter, I examine Franciscan discussions of the natural world. Remarking on the Franciscan pursuit of alchemy, Lynn Thorndike offered this gem: "The recording angel must smile frequently at the little ironies of history. One of these amusing inconsistencies of real life is that followers of St. Francis, the apostle of poverty, should have interested themselves in making gold."[27] While the accumulation of wealth would certainly have appeared unseemly to critics of the Order, the pursuit of alchemy in and of itself was in fact consistent with the Order's interest in the natural world. Multiple generations of Franciscan authors, including the Order's founder, are deeply invested in the celebration and investigation of the natural world. As Francis's *Canticle to Brother Sun* illustrates, the earth and the very elements that formed it were a *donum Dei*, a gift of God. The Order's founder saw no conflict between the spiritual life and the material world. Later commenters would turn the focus of Francis's devotion to a more considered treatment of natural philosophy. Bonaventure and Peter Olivi in particular argued that natural philosophy was essential to plumbing the spiritual value of the cosmos. Olivi's adherence to a literal understanding of Genesis only reinforced the pursuit of natural philosophy as a necessity. Therefore, however marginal the study of alchemy was to Franciscan intellectual life, it was hardly out of step with general intellectual currents within the Order.

The second chapter turns to the elixir tradition proper. Tracing the genealogy of various elixirs underscores the centrality of cosmological traditions and assumptions in the formulation of this cure-all. Though each author maintains an adherence to a broad Aristotelian concept of alchemy and natural philosophy, Christian theology and canon law are critical to the development of specific iterations of the elixir. Roger Bacon models the elixir's powers on the Christian concept of the resurrected body. The alchemy ascribed to Vitalis of Furno appears to be shaped narrowly to fall within the confines of what is licit according to the Franciscan Order and the papal curia. John of Rupescissa considers his elixir to be a literal distillation of heaven. While each substance is a cure-all, there is great diversity in the nature of the compounds, not all of which can be explained solely by resort to alchemical traditions. Religion may constrain or inspire the alchemist, but it is never incidental.

In the third chapter, I take up the role of apocalypticism in relation to alchemy. The Franciscan Order was thoroughly embroiled in apocalyptic speculation during the thirteenth and fourteenth centuries. As intramural disputes over poverty and the legacy of Francis heightened apocalyptic rhetoric, those brothers not consumed with the approach of the end times still had to deal with those who were. Apocalyptic speculation was part of the oeuvre of each of the alchemical authors under consideration, and, as such, I treat it independently of other religious ideas. Given the few alchemical texts emerging from the Order in the thirteenth and fourteenth centuries, it would be foolish to suggest that Franciscan theology or apocalyptic thought generated alchemical thought. Yet, on their own both alchemy and apocalyptic thought offered the possibility of realizing the spiritual in the material world. Taken together, these practices could be mutually reinforcing.

Chapter 4 takes on the issue of Franciscan religious practice. While there is a tradition of alchemy as a religious practice in late antiquity, I argue that the devotional aspects of Bacon's and Rupescissa's elixirs are better approached through the lens of Franciscan ritual life. Franciscans, like all regular orders, marked their days, years, and lives liturgically. I discuss briefly how these liturgies opened up a means of recontextualizing the relationship to God and the cosmos. While I do not argue that alchemy was a ritual per se, the production of the elixir seems to be enhanced when the alchemist enters the ritual mode. The devotional approach to alchemy opened new avenues for these medieval theorists to interpret their practice and their results.

CHAPTER 1

Franciscans and the Sacral Cosmos (The Context of Franciscan Alchemy)

Praise be to you, my Lord, and to all your creation. . . .
—Francis of Assisi, *Canticle of Brother Sun* (1224/5)

The goal of this chapter is to illustrate in brief Franciscan considerations of the natural world. This discussion is more than just a backdrop to the elixir. It also provides a sense of how and why Franciscan discussion of the elixir became entwined with religious ideas and language. The discussion that follows complements, but also offers important context to, the arguments that have been made about the religious turn in alchemy in the fourteenth century.[1] The religious turn must be considered in light not just of changes in alchemical discourses, but also of corresponding developments in theology and devotional practices. Religion and religious language emerged in Franciscan alchemy because friars were already busy drawing theological conclusions from nature and philosophical conclusions from scripture. Unsurprisingly, perhaps, for an order that eschewed monastic enclosure, the friars embraced the created—that is to say, the material—world.

In spite of the variety of intellectual approaches taken toward the natural world, the principal concern of Franciscan writers—and likely of all clerical scholars of the era—was not with nature per se, but with *creation*. *Natura* in any case did not mean nature in quite the way we mean it today. Medievals had no notions of ecology or ecosystems. Rather *natura*, when considered as an intellectual category, tended to follow Hugh of St. Victor's influential definition.[2] Hugh considered a tripartite division between inclinations or properties of a specific thing or of a group of things. Hence, horses have a nature that is common to them that explains horse-like behavior. Likewise, a specific horse has its own nature—fast, slow, temperamental, calm—in comparison to other horses. Hugh also notes, however, that sometimes *natura* was used to mean something more like the natural order.

It is this latter definition that is pertinent to the discussion in this chapter, and is the one that arrives closest to nature as a kind of cosmological category. The natural order was inextricably linked to the medieval understanding of the creation of the world as outlined in Genesis. "Regarding corporeal nature, namely how it came to be," writes Bonaventure, "we must comprehend that it was produced in six days, such that in the beginning before any day, God created heaven and earth."[3] There is no substantive divide, then, between the natural order of things and the medieval notion of creation.

The link between God and nature is something of a truism for medieval writers. Franciscans, however, beginning with Francis himself, did not limit themselves to a kind of standard preamble of nature's ontological link to God before engaging in Aristotelian speculation. Nature was not a neutral philosophical category. Instead, Franciscans developed sacral ideas of creation. Their philosophical discussions of the natural world were not just inflected by Christian religion, but aimed at understanding the role of the created world in salvation. At times mystical, at times manifestly Aristotelian, Franciscans' ruminations on the universe probed the nature of God's relationship to humankind and the human relationship to the rest of the cosmos.

Much scholarly discussion, in the Middle Ages as well as lately, has focused on the rigors of Franciscan poverty and humility, and the ensuing divisions fomented by various interpretations of Francis's commands and practices.[4] For all its ascetic tendencies, however, Franciscan piety was not fueled by mistrust of the material world. Franciscans certainly drew—as did most of the religious of the Middle Ages—on the eremitical tradition handed down from the desert fathers, but the poverty and spiritual discipline practiced by the friars should not be mistaken for dualism or a fundamental suspicion of material things. Sin, argued Francis, was a result of the will, not a result of merely inhabiting the world.[5] Hence Franciscans, like other orders of friars, eschewed the rhetoric and practice of enclosure and engaged with society and the world around them. Even more telling, Francis's own writings focused more on traditional Christian practices of *material* piety, such as care and reverence for the Eucharist, than they did on poverty, so often seen as central to the Franciscan ideal.[6]

This Franciscan focus on the natural world and material aspects of Christian practice also emerged in the context of the Church's conflict with heretics, both real and imagined. While the facticity and prevalence of heretics has been the subject of scholarly debate for decades, that the Church *perceived* a threat is not in doubt.[7] Especially, though not exclusively, Cathars offered alternative theologies of the material world that repudiated its essential goodness. Christians were no strangers to the notions that the physical body presented a hurdle to salvation and that the world possessed a myriad of dangers and temptations, but the promised Resurrection provided a final exclamation point to the idea that God had created people and the world as fundamentally good. Saint Francis himself stressed that the root of sin was found in the human heart—a flaw of the postlapsarian individual, rather than of the postlapsarian world. Cathars, however, believed that the physical world was a mistake, created by the Devil or a flawed demiurge and hence inextricably linked to sin. Only the shedding of the material world offered salvation. The evil of the body and the material world was not limited to sexual sins, but to materiality more generally. Cathar *perfecti,* the "Good men" and "Good women" who made up the supposed Cathar prelacy, had to refrain as much as possible from dirtying themselves with the physical world. Hence they refrained—at least nominally—from sex, eating meat or other "products of coition" (this left fish as a possible meal, since they were thought to be generated spontaneously from flowing water), as well as other more standard sins. Any deviation from this path resulted in the loss of their saved status.[8]

Again, whether or not Cathars actually practiced or preached dogmatically rigid dualism, it was the clear understanding of the Catholic Church that they did. The anti-material doctrine and austere lives of Cathar perfecti resonated with the eremetical traditions of Christianity only recently revived in the eleventh and twelfth centuries. And, regardless of how many perfecti may have been wandering the Languedoc, the Rhineland, or northern Italy, their spare frames and poor mien offered an implicit critique of the Latin Church. The Franciscans, much like the Dominicans, were in the vanguard of opposition to these heretical ideas, and it is in the context of such an imagined battleground that Franciscan spirituality and exhortation arose. Proving the essential goodness of creation, especially via preaching, was a key strategy in reinforcing orthodoxy.[9] Therefore, didactic material on nature produced by the Order—everything from Francis's own teachings and poetry, to the legends of his life, to encyclopedia entries—was at some level apologetic. Yet, none of this was simply rhetoric either. Rather, it is more accurate to say that the teaching of the Church on the inherent goodness of creation resonated deeply with Francis and his followers.

What follows is a deeper exploration of Franciscan views on the cosmos. The sacral orientation of Franciscan writing on nature is a crucial context for understanding Franciscan alchemical writing, serving as a link between alchemy as found in the Franciscan encyclopedias and the seemingly more radical speculation of Joachite-inspired alchemists, such as John of Rupescissa and Roger Bacon. Indeed, the latter writers, while long considered central to the development of medieval alchemy, have frequently been seen as outliers among their Order, thanks to their alchemical speculation. The discussion that follows demonstrates that while alchemy might have been a marginal practice, it is deeply tied to broader currents of Franciscan intellectual life and, as importantly, sheds some light on more mainstream Franciscan considerations of the material world.

The Natural and the Mystical: Francis and Bonaventure

The identity of no religious order was so tightly intertwined with the life of its founder as that of the Friars Minor. Christendom's greatest saint—no other held the title of *alter Christus*—Francis held an undeniable sway over the Franciscan imagination both before and after his death. The standard set by Francis's humility, poverty, and charity was likely an impossible one to live up to collectively. And, as debates over the role of the friars regarding adherence to apostolic poverty and involvement in elite learned culture evolved over the thirteenth and fourteenth centuries, interpretations of the life and teachings of Francis were fraught with apologetic intent. Therefore, while the various lives of Francis include a rather substantial amount of material on the saint's relationship to nature, it is best to begin with Francis's own writings. For, in spite of Francis's reputation as a lover and friend of nature, only one authentic work of his on this topic survives: the *Canticle of Brother Sun*.

The *Canticle* is likely the earliest Italian poem to survive into the present day, and scholars have long noted with little controversy that it draws much of its structure from Christian liturgy. Its structure and call to prayer strongly resemble Psalm 148, the *Laudate*, which Francis, like other religious, recited every morning.[10] There are questions about how much Francis may have been influenced by troubadour poetry, but he did mean the Canticle to be sung by Franciscans as troubadours of Christ. Indeed, its first public performance appears to have facilitated a rapprochement of the feuding secular and clerical factions in Assisi.[11] Leaving aside questions of what influenced Francis stylistically, let us take a look at the Canticle itself. It is relatively short, so I have quoted it here in full.[12]

Altissimu, onnipotente, bon Signore,
Tue so' laude, la gloria e l'honore et onne benedizione.

Ad Te solo, Altissimo, se konfane,
et nullu homo ène dignu Te mentovare.

Laudato si, mi' Signore, cum tutte le Tue creature,
Spezialmente messor lo frate Sole,
lo qual è iorno, et allumini noi per lui.
Et ellu è bellu e radiante cum grande splendore:
de Te, Altissimo, porta significazione.

Laudato si', mi' Signore, per sora Luna e le stelle:
in celu l'ài formate clarite et pretiose et belle.

Laudato si', mi' Signore, per frate Vento
e per aere et nubilo e sereno et onne tempo,
per lo quale a le tue creature dai sustentamento.

Laudato si', mi' Signore, per sor'Aqua,
la quale è multo utile et humile et pretiosa et casta.

Laudato si', mi' Signore, per frate Focu,
per lo quale ennallumini la notte,
et ello è bello et iocundo e robustoso e forte.

Laudato si', mi' Signore, per sora nostra matre Terra,
la quale ne sustenta e governa,
e produce diversi fructi con coloriti flori et herba.

Laudato si', mi' Signore, per quelli ke perdonano per lo tuo amore
e sostengo infirmitate et tribulatione.

Beati quelli ke 'l sosterrano in pace, ka
da te, Altissimo, sirano incoronati.

Laudato si' mi' Signore per sora nostra Morte corporale,
da la quale nullu homo vivente pò skappare:
guai a quelli ke morrano ne le peccata mortali;
beati quelli ke trovarà ne le tue santissime voluntati,
ka la morte secunda no 'l farrà male.

Laudate et benedicete mi' Signore' e ringratiate
e serviateli cum grande humilitate.

Most high, almighty, good Lord,
Praise, glory, and honor, and all blessings are yours.

To you alone, most high, do they belong,
And no man is worthy is to speak your name.

Praise be to you, my Lord, and to all your creation,
Especially Sir Brother Sun,
Who is our day, and you give us light through him.
And he is beautiful, shining with great splendor.
From you, most high, he takes his meaning.

Praise be to you, my Lord, from Sister Moon and the stars:
In the heavens you have formed them, shining and precious and beautiful.

Praise be to you, my Lord, from Brother Wind,
And from the air and cloud and calm and all weathers,
Through which you give your creatures nourishment.

Praise be to you, my Lord, from Sister Water,
Who is so useful and humble and precious and chaste.

Praise be to you, my Lord, from Brother Fire,
Through whom you lighten our night:
And he is handsome and merry and vigorous and strong.

Praise be to you, my Lord, from our sister Mother Earth,
Who nourishes and sustains us,
And brings forth her various fruits, with many-colored flowers and grasses.

Praise be to you, my Lord, from those who forgive for love of you,
Bearing illness and tribulation.

Blessed are those who bear them in peace,
For by you, most high, will they be crowned.

Praise be to you, my Lord, from our Sister bodily Death,
From which no living man can escape:
Woe to those who die in mortal sin:
Blessed are those whom she shall find doing your
most holy will,
For the second death shall not harm them.

Praise and bless my Lord, and give him thanks
And may you serve him with great humility.

The central debate over the meaning of the poem—a debate that reaches back to the Middle Ages—is how to interpret the Italian *per* first found in line 10: "*Laudato si, mi signore, per sora luna e le stelle.*" The issue is that the term *per* can be translated as either "for" or "by," depending on whether one reads the term as closer to the Latin *propter* or the French *par.*[13] So, one could read this line as either commanding that God be praised *for* his creations, Sister Moon and the stars or, more controversially, *by* duly anthropomorphized celestial bodies. One could also see *per* as meaning *through*. There are pros and cons to each translation, and it is fully possible that Francis may have intended the passage to be somewhat ambiguous. After all, his use of the Umbrian dialect rather than Latin is quite deliberate. Though Francis's works are few and usually brief, only one other piece, a Canticle composed for the Poor Clares the same year (1225), is in Italian, or, more properly speaking, the Umbrian dialect. The rest of his writings are in Latin.[14]

The issue of how to translate *per* bears on more than just linguistics. It lends weight to various interpretive arguments. If we take what I would consider to be the narrow reading of *per* as "for," then the theme of the Canticle is rather simple—a praise to God for creation. This more conservative reading has been favored to varying degrees by Thomas Nairn, O.F.M, André Vauchez, and Augustine Thompson, O.P.[15] These scholars emphasize this reading as fitting the Middle Ages and as an important corrective to earlier scholarly claims that Francis expressed a kind of pantheism in the poem. Written while Francis was dying and nearly blind, the poem stresses the goodness of creation and humankind's duty to praise God for such a gift.[16] Michael Blastic likens the poem, appropriately, I think, to a sermon, and Vauchez has called it a Mass for the world.[17] The argument in favor of this conservative translation is that the Canticle's simplicity should be considered not a slight to Francis's work, but rather a sign of its depth and power.

Other readers have favored, however, reading *per* as "by," which significantly alters the meaning of the poem. Roger Sorrell in the 1980s did much to popularize this view in Anglophone scholarship, but his argument has been undercut by his claims that Francis was a nature mystic, and by (to be frank) somewhat unfair readings of his work, which have overlooked much of the nuance of his discussion of embedded mysticism within the poem.[18] Perhaps the best argument to be put forward recently for interpreting *per* as by or through has been made by the Italianist Brian Moloney. To be clear, Moloney favors a more ambiguous reading of the poem, but offers a fairly strong defense for why one could consider *per* as "by" instead of "for."[19] At the beginning of the poem, the speaker says that he is not worthy to say the name of God. If this is so, then it begs the question of who exactly is doing the praising. There is precedent here in Francis's first Rule—one that was never papally approved—where he invokes Christ and the Holy Spirit to give thanks to God since he is not worthy to speak God's name. This is, ultimately, an Augustinian claim, and not altogether new. Likewise, in Psalm 148, on which the Canticle is modeled, the Vulgate text uses the imperative mood to command creation to praise God. Perhaps most important, this interpretation is precisely what Francis's immediate followers *thought he meant.* Both the compiler of the very early Assisi compilation—stories about Francis that were gathered to be used in his official *Life*—and no less a personage than Saint Bonaventure believed the Canticle was a paean *by* creation, not for it. Bonaventure went so far as to replace the preposition *per* with the much more straightforward *da*, to emphasize that praise was to be given by creation.

This latter reading of the Canticle, as I have mentioned, has been related to readings of the Canticle as an expression of Francis's mysticism. Moloney points out a key passage from one of Celano's lives of Francis that supports such a conclusion: "Often as he walked along a road, thinking and singing of Jesus, he would forget his destination and start inviting all the elements to praise Jesus."[20] Here we see a strong connection to the Canticle, where beginning in line 10 and running through line 22, the elements are specifically called on to praise their Creator. In the passage from I Celano, however, Francis's prayers are Christocentric. This is fitting with much of what Francis has left us in his writings, where the Son is the principal person of the Trinity whom Francis engages, contemplates, and remembers. Yet, toward the end of his life, his writings shift from an emphasis on Christ to the Father, which fits the theme of the Canticle.[21] In the Canticle it is the Father, the Lord, who is the principal object of praise.

We do need to be cautious, however, regarding Thomas's description of Francis's enraptured prayers as definitive. Thomas's description of Francis could have drawn on traditional Augustinian and pseudo-Dionysian themes of mystical experience. In this formulation, one's focus is so entirely upon God that he or she is alienated from earthly senses. In both the first and second life, when Thomas also describes Francis as spiritually absent from earth during some of his contemplations of God, he seems to portraying Francis in ways that fit with this tradition. It is not clear whether Francis knew of such descriptions, but Thomas presumably did. This is not to say that Francis was not lost in contemplation, but the hallmark of pseudo-Dionysian mysticism is alienation from the world, not its celebration.

Parallel descriptions in Bonaventure's *legenda maior* are even more suspect, as Bonaventure's own mystical works were deeply influenced by pseudo-Dionysius. Bonaventure clearly regarded alienation from the world as congruent with mystical union. Indeed, in the last chapter of his *Journey of the Mind to God*, Bonaventure explicitly cites Francis as having developed disaffection from the senses and the world to the point that he would have appeared outwardly dead (*quasi exterius mortuus*) while rapt in mystical contemplation (*in excessu contemplationis*).[22] Though we will get to Bonaventure's views on nature shortly, it can be said for the moment that Bonaventure agrees God is reflected in the creation—"Anyone who is not illuminated by the splendor of created things is blind," he states. Alienation therefore is not the same as rejection: "Created things of every kind in this sensible world signify the invisible things of God, mainly because God is the origin of every creature, its exemplar and end."[23] Bonaventure considers contemplation of creation to be a lower rung of the ladder of mystical speculation, but a necessary one. If we are to consider the Canticle as a kind of mystical poem, or the result of mystical contemplation of God, the progression from contemplation of creation to celebration of God is much more direct and much shorter than Bonaventure's intellectual scheme.[24] I am not convinced that formal mysticism should be attached to the Canticle, but if it is, it speaks more to a kind of passionate, immediate experience of God in nature that is particular to Francis and an experience that needs to be appropriated or at least mediated by Bonaventure to fit within a more conventional understanding of mystical union.

I want to return now to the description of creation in the Canticle. In lines 10–22, Francis invokes the four elements—air/wind, fire, water, and earth—to praise God. Most scholars describe this passage as a depiction of the Aristotelian elements, but such a distinction implies a more philosophical stance than is likely present. Francis was not calling on the elements as a sort of proto-periodic table. The import of the four elements as building blocks of the universe was more foundational—more of a truism than anything else.[25] Naming the elements does not necessarily imply philosophical contemplation of them. Modern analogies are hard to come by, but it is something like saying the earth is part of the solar system—a common fact most people assume to be true whether or not they know anything about astronomy.

Francis invokes these elements immediately after the heavenly bodies. From the twelfth century onward the view became prevalent that the heavenly bodies were not fiery bodies but rather concentrations of a fifth element—an unchangeable and perfect element fitting of heaven, whose luminosity was the result of relative density.[26] If Francis was familiar with this view, which circulated during this period though not axiomatically like the rest of the four elements, then it could be possible that he saw nature in what was, in that time, a scientific manner. The evidence for such a view is rather tenuous. Some scholars believe Francis was familiar, through Brother Elias, with the basic tenets of alchemy, where many adherents did consider a fifth essence to be part and parcel of the created world.[27] Evidence for Elias's alchemical activity is sparse and indirect, however, and Francis certainly leaves no direct trace of such knowledge anywhere in his works.[28] In any case, later friars themselves disagreed on the question of the fifth element. Bonaventure, in his thirteenth collation on the six days of creation, says that no one knows the truth about the substance of the heavenly bodies, though in earlier works, such as his *Sentence Commentary* and his *Reduction of Arts to Theology*, he assumes that "there are four elements and a fifth essence," all of which can be grasped by human senses.[29] Bonaventure's contemporary Roger Bacon outlined in a variety of works an almost unquestioned assumption that the heavens were made of a fifth essence. This latter view seems to have won out (probably not thanks to Bacon), and by the early fourteenth century the biblical commentator Nicholas of Lyra can include it without justification in his discussion of creation.[30]

The logic of the Canticle, however, makes more sense if we consider it less as a philosophic statement on the universe than as a circular progression from the personal to the cosmic to the personal. After the initial laud given by the speaker of the poem come the heavens—the cosmos around the earth—and then the focus descends to the world itself (the four elements). The next verse is something of an anomaly. It was added later, and there is neither Sister nor Brother as its subject. Finally we move to Sister Death, the end of worldly things, who deserves a personal welcome from each of us. This is a difficult verse, since biblical sources frequently treat death as something to be conquered and hold that death is not the creation of God. It is important to remember, however, that Francis knew well he was close to death at the time of the composition of the Canticle and, I think, wanted to express that the nearness of death should not dampen one's admiration for creation and for God. Thus, the last verses deal with human experience.

While the poem is far from a theological statement on creation, nevertheless a certain theology underpins it, namely the familial treatment of creation. Francis called members of his Order Brothers (and the Poor Clares, Sisters), and made a point of rejecting his birth family for his spiritual one. Francis clearly wanted to convey a sense of community, even family, with creation. There is a tension, then, in the poem between what I would describe as a vertical and a horizontal relationship with creation. The vertical relationship is typical of the medieval period and is expressed in how the elements (or plants, trees, and various animals) are useful to humanity. The horizontal is the more radical aspect and expresses the familial aspect of creation—humanity is no more or less a creature than anything else.

This tension is what unites the Canticle with other aspects of Francis's relationship with nature. The horizontal relationship to nature is frequently expressed in stories of Francis and animals. The sermon to the birds, for instance, though problematic in its many versions, expresses this kinship, as do Francis's encounters with other animals. At the same time, he did have his difficulties with animals—he saw mice as tools of the devil and called one companion Brother Fly as a rebuke to his laziness. Moreover, Francis was hardly a vegetarian and, somewhat unusually, allowed his brothers to eat meat. And while he frowned on the riding of horses by brothers, he did so not out of compassion for the animal, but because such actions did not befit the humility of the Friars Minor.[31] Thus, Francis embraced the traditional vertical medieval view, but also incorporated a very personal, intense, and direct relationship with creation that stressed universality and community, where all things had their origin in God.

Francis's concern for nature was, in the Middle Ages as it is today, one of his defining characteristics. Neslihan Şenocak has determined that Francis's special relationship to the natural world was one of the few things many of the first couple of generations of brothers actually knew about their founder. Materials used to generate Francis's life were not collected until 1244, and generally speaking the first brothers outside central Italy knew very little about Francis himself, outside his reputation for holiness.[32] Yet one of the first brothers to reside at the nascent convent near the University of Paris, Julian of Speyer, was keen to point out Francis's affinity for the created world:

> What do you think he drank in of true knowledge, sweetness and grace in the sun, the moon, the stars and the firmament, in the elements and in their effects or embellishments? What, I ask, did he drink in when he contemplated the power, the wisdom and goodness of the Creator of all in all things? Surely, I do not think that it would be possible for any mortal to express this in words. Since he traced all things back to their one first beginning, he called every creature "brother," and, in his own praises, continuously invited all creatures to praise their one common creator.[33]

Whether or not the *Canticle* was copied down or sung, it is at least clear that friars in the academic milieu of Paris were familiar with Francis's special rapport with nature. It was likely, however, that most friars understood this to be a personal characteristic rather than part of the Franciscan identity—Francis's own works seem to bolster this idea, since traditional Franciscan topics such as humility, obedience, and poverty are the cornerstones of his own writings. It is also significant that in Bonaventure's *legenda maior*, which came to be the definitive life of Francis, Francis's horizontal relationship with animals is subtly transformed into what Augustine Thompson has characterized as the prototypical "white magic" of the saints, where Francis commands nature as much as he finds fellowship in it.[34] André Vauchez believes that much of this reinterpretation of Francis by Bonaventure was deliberate. Bonaventure was seeking to forestall the rending of his order by the pull between imitating Francis and serving the needs of the Church and the papacy.[35] Bonaventure does not, however, impugn the importance of nature, and in fact comments on it frequently in his works. As a follower of Francis and a mystic himself, Bonaventure was deeply moved by Francis's connection to the created world. And, as a theologian, Bonaventure linked the encounter with the world to more traditional avenues of thinking about the created universe.

Genesis and Natural Philosophy

The notion that the opening chapters of Genesis constituted a text of natural philosophy is recognizable at the beginning of the scholastic era. Masters of cathedral schools such as William of Conches and Thierry of Chartres considered the cosmology of Genesis from a philosophic standpoint. Peter Abelard, in his commentary on the Hexaemeron, to take another example, begins his gloss of Genesis 1:1 (*In the beginning God created heaven and earth*) by stating that this passage refers to the creation of the four elements, which are themselves the basis of all material created things.[36] Written before the widespread dissemination of the Aristotelian corpus, Abelard's discussion of Genesis relies on the Platonic cosmology discussed in *Timaeus*. For example, he considers heaven properly speaking to be composed of fire, rather than of a fifth, immiscible element, and later refers to these initially created elements as *hyle*, or prime matter.

While the discussion of heaven or *aether* as fire is cause for some mental gymnastics when discussing the separation of the waters (Gen 1:6), this in itself is a notable point. Rather than spiritualizing the text of Genesis to skip over what appears to be a plain contradiction to Plato's cosmology, Abelard goes to great length to explain and synthesize.[37] This is rather a common refrain in later Franciscan commentaries as well. Like Abelard, who relegates the spiritual value of Genesis to a brief passage at the outset of his work, Franciscan exegetes concentrate almost exclusively on the natural philosophic or cosmological implications of the six-day work. Abelard also understood Genesis as a jumping-off point for discussions of practical philosophic elements. Abelard does not discuss alchemy—the alchemical corpus translated and adapted from Muslim scholars were not yet fully diffused in Latin Christendom—but Abelard does connect the six-day work to the practice of astrology.[38] This leap from biblical to philosophical is made natural by Abelard's offhand comment that Moses, according to tradition the author of Genesis, learned astronomy among the Egyptians.[39] The implication of such a comment is that Moses knew full well he was writing a philosophical text when he penned Genesis and intended it to be understood precisely in the way Abelard wrote about it.

Latin Christians inserted Genesis into discussions of alchemy almost as soon as they began to compose original treatises. In the 1257 *Liber seceretorum alchimie*, Constantine of Pisa (or the master who taught him) took pains to adapt the Neoplatonic heritage of alchemy to the manifest reality of the truth of Genesis.[40] Though not a clerical author per se, Constantine is pressed to make Genesis present in his work. As the word of God, the book of Genesis presented an authoritative view of the world. This view in turn colonized and rewrought surviving Neoplatonic and Aristotelian theories.

Genesis, like any text, derived its meaning as much from the context in which it was read as from its content. Among Franciscan scholastics the Hexaemeron was certainly a font of spiritual wisdom, but such wisdom came from meditation on the nature of the material world described therein. Just as was the case in the meditations of Francis, spiritual value flowed from meditation on the material. The physics of creation was not an ancillary aspect of Genesis. Rather, the nature of the material world undergirded spiritual assumptions and religious truths. Assumptions about the physical composition of the world were imbricated with assumptions about God, and humanity's relationship to Him. Neither the physical nor the spiritual could sidestep the other.

Bonaventure's Book of the World

A century after Peter Abelard, Bonaventure approached the story of Genesis in a rather different manner in his *Breviloquium*. Bonaventure composed the *Breviloquium* the year (1257) he left his academic post at the University of Paris to take on the role of Minister General of the Franciscan Order. Unlike the *Collationes in Hexaemeron*, which approaches creation almost entirely symbolically, the *Breviloquium* addresses scripture as "partly plain speech, partly mystical (*partim plana verba, partim mystica*)."[41] Like the *Summa* produced by Aquinas and other like texts spawned by Bonaventure's contemporaries in the universities, Bonaventure idealized the *Breviloquium* as an introductory theological text that was, as Dominic Monti has said, a "coherent synthesis" of critical theological issues. Yet, as a text that achieved (and not just aspired to) brevity and that used deductive logic, the *Breviloquium* was not typical of the theological treatises of its era. Bonaventure forewent the *quaestio* method he employed in his *Sentence Commentary*, and penned instead a cogent and authoritative treatise that invited neither debate nor argument.[42]

While the *Breviloquium* is not a work of exegesis per se, the authority of Bonaventure's authorial voice emerges from his assumption that he is presenting the essential wisdom of Christian scripture, rather than engaging in a kind of specific philosophical or theological inquiry. Scripture, Bonaventure points out in his prologue, is not restricted (*coarctatus*) by reason or philosophical condition. Rather, it deals with every aspect of the universe pertinent to salvation.[43] While modern readers might see salvation as a limiting factor to philosophical discussions, this was not the case among Bonaventure's contemporaries—at least in theory. Bonaventure elsewhere was quite caustic in regard to philosophical speculation, especially Aristotelian philosophical speculation, which had no spiritual aims, though he was happy to employ Aristotelian logic and principles when useful.

Fortunately for this discussion, the creation of the physical world did pertain to salvation, and hence makes up the subject of the second part of the *Breviloquium*. After dealing succinctly with any possible opinions contrary to the notion that God made the universe from nothing (an argument against Aristotle's claim that the cosmos was eternal), Bonaventure proceeds to outline the six days, listing the things created on each (day 1, light; day 2, firmament; day 3, separation of the waters to make land; day 4, heaven adorned with celestial bodies; day 5, air and water filled with birds and fish respectively; day 6, the land gets animals and human beings). Bonaventure is keen to note that the six days allow for many groupings of threes, which he regards as a sign of the Trinity.[44] The religious symbolism, however, does not detract from his focus on the physical description of the cosmos, but rather relies on it. "The creation of the world is like a kind of book, in which the Trinity, its artificer, shines forth, is represented, and is read."[45] Bonaventure regards the world not merely as a creation to be praised, but as a manifestation of God.[46]

In the *Breviloquium*, Bonaventure reads the book of the world quite literally. Having detailed the six days, Bonaventure describes the present world schematically, by creating a series of subdivisions. The first division is between the heavenly (*caelesti*) and the terrestrial (*elementari*). Heaven he subdivides into, not surprisingly, "three principal (*tres principales*)" categories, the empyrean, the crystalline, and the firmament. This does not include the starry heaven, which is somewhat awkwardly tacked on to the "principal" categories, until he promptly divides it into seven parts for the seven planets. Therefore, when Bonaventure later states that the world is perfectly ordered, part of his argument rests on the implicit correspondence between significant Christian numbers and the observable world. Bonaventure also notes the traditional four elements as terrestrial spheres. His discussion of the elements is interesting, for even though the elements are changeable and prone to degradation, Bonaventure notes that they too are part of the perfect order as they allow for a diversity of forms to exist.[47]

Bonaventure is particularly concerned about the relationship between heaven and earth, which he understands to be mediated by the planets. Therefore, like Peter Abelard before him, Bonaventure uses his reading of Genesis to make some statements about astrology. While Bonaventure objects to the concept of fate, he does hold that the seven planets correspond to the three heavens and the four elements.[48] The effect of celestial bodies on the terrestrial world is his focus here. Celestial influences, according to Bonaventure, affect "the production of generable and of corruptible things, such as minerals, plants, sentient creatures and human beings."[49] The relationship between the celestial and the terrestrial will be discussed in much more detail in the following chapter, but here it demonstrates Bonaventure's attention to the "facts" of the observable world, as well as to the meaning of the structure of the cosmos.

While it might seem that Bonaventure is traveling fairly far afield from Genesis, it is quite clear that he understands Genesis to lead to and perhaps require a natural philosophic analysis. Reading the Book of the World to find the vestiges of God requires examining the world in detail, not simply overlooking the physical in favor of the spiritual. Perhaps sensing this objection, he turns in the fifth chapter of the *Breviloquium* to a discussion of how we arrive at a discussion of natural philosophy even if the Bible itself does not tell us about "the distinction of the spheres, nor the heavens, nor the elements."[50] While he goes on at some length to answer the question, the basic upshot is that scripture ought to be understood in comparison to the Book of the World. Scripture, for Bonaventure, is about the revelation of Christ and his teaching. It is redemptive or repairing. The Book of the World, namely creation, is a book about the Father, the Creator. Thus, scripture includes only what readers *need* to know about redemption, and should not be seen as somehow closing off knowledge of the world. Scripture is sufficient, but not all-encompassing.[51]

The Book of the World, however, is mostly closed, according to Bonaventure. While the order of the created world might once potentially have led human beings to the divine (as it did in the Garden), postlapsarian human beings are not so fortunate.[52] In his *Collations on the Six Days*, a treatise where creation serves a metaphorical purpose, Bonaventure is quite clear that people cannot simply observe nature and know God, that is, there is no sense of any kind of natural religion.[53] Bonaventure argues, then, that the role of scripture is to begin to repair what was lost to human beings at the fall, namely, their "eye of contemplation (*oculum contemplationis*)" as Hugh of St. Victor called it.[54] God endowed humanity at creation with an eye of flesh, an eye of reason, and an eye of contemplation. The eye of flesh allows people to see around them; this is the literal function of an eye. The eye of reason is what allows one to see inside oneself, that is, to think. The eye of contemplation, however, is what allows the other kinds of eyes to look heavenward in an anagogical sense. With the contemplative faculty damaged after the fall, the physical world can be seen, described, catalogued, and theorized, but it does not ultimately lead, as it should, to the manifestation of God.[55] Through recalling the beauty and goodness of God, scripture (along with grace) aids in the repair of the senses, so that the science of creation reveals its author.[56]

What Bonaventure adds to the Franciscan understanding of nature is a sense that the kind of immediate sacrality felt by Francis in his apprehension of God in creation can also emerge through reason, philosophy, and theology. It is implied that the saintly Francis was able to read the Book of the World, and immediately see its truth. Bonaventure's path is more deliberate, but it embraces every kind of knowledge about the natural world that leads to the divine. The four elements and three heavens are not biblical in origin, yet to the Christian who also reads the scriptures, they can be seen to refer to the seven days of creation, or (in the case of the heavens) to the Trinity. That said, natural philosophy is not an end in itself. Even when it leads to knowledge of the Creator, it is not the highest kind of knowledge. Bonaventure reserves that for the kind of mystical union discussed in the earlier part of this chapter. Bonaventure's distinct celebration of the natural world, however, would resonate in the next generation of Franciscan scholars. We turn now to one of those, Peter Olivi.

Creation as a Mirror of the Trinity: Peter of John Olivi

Peter of John Olivi's reputation has grown recently, thanks to a great deal of scholarly work that has more firmly established the context of his life and thought. In particular, scholars such as David Burr, David Flood, O.F.M., and Kevin Madigan have read Olivi in the context of his era rather than through the lens of later Spiritual Franciscans and *fraticelli* who appropriated his views. Olivi is now considered one of the premier intellects of the Franciscan Order at the end of the thirteenth century. The marginalization of his ideas was a product of historical contingency rather than a reflection on their relative merit. Olivi's allegiance to the exegetical methods (and apocalyptic fears and hopes) of twelfth-century abbot Joachim of Fiore certainly had much to do with opposition to his ideas in his own period, but he was far from the only Franciscan to suffer on account of his attachment to Joachism. The apocalyptic strains of Joachism will be tackled in more depth in Chapter 3, but to understand Olivi's views on creation, it is necessary to outline a few of the salient details of Joachim's mode of exegesis which Olivi adopted.

Discussion of Joachite exegesis often highlights its more radical and apocalyptic features. Olivi's fidelity to this exegetical strategy, however, displays the more subtle aspects of its inventiveness, especially the principle of *concordia.* Like traditional typological readings of scripture, *concordia* argues for intertestamental links. It does not, however, stop there, as the intertestamental relationships could also describe future events. According to Joachim, the two testaments are a complete work of scripture that cannot be superseded, but they record a history of an age of the Father (identified with the Hebrew Bible or Old Testament), and the beginning of an age of the Son. By understanding the parallels between the two testaments one can use them to intrepret history since the age of the apostles and thus deduce the structure of the remaining age of the Son and the third age of the Holy Spirit. *Concordia* highlights the numerical patterns of the Bible as particularly instructive, as they can be found in a wide variety of places. We saw this to a degree in Bonaventure, who adopted some of the exegetical strategies of Joachim, but opposed the Joachite idea that the books of the Bible could be used to see the order of the future.[57] Olivi, however, is sympathetic to this view, and in his commentary on Genesis puts special emphasis on patterns of three that mirror the Trinity. In the prologue to the commentary, Olivi puts *concordia* front and center. Again, the principle of *concordia* relies on earlier biblical typology, and its novel aspects are not always immediately obvious. For instance, citing Genesis 18:4–5, when Abraham greets the three angels, Olivi tells his readers that these three angels signify the three persons of God.[58] This Trinitarian formulation is at the heart of Olivi's Genesis commentary, but was already a longstanding typological interpretation. Yet, in the friar's view, there is no aspect of creation that fails to reveal the Trinity and enrich one's understanding of it.

Genesis, Olivi tells us, is like a tree. Its roots are the six days, its trunk is the line of patriarchs, and its branches are their descendants, divided into the twelve tribes of Israel. Its leaves are the laws, and its flowers the wisdom literature and the prophets.[59] Again, Olivi's tree metaphor is not new. The tree long served exegetical thinkers as a means of mapping the genealogies beginning in Genesis (and concluded in the Gospels). Yet, for Olivi the story of Genesis does not—and should not—stand on its own. "Under the tree," he says, "is the entire Trinity of God, the threefold hierarchy of angels, and *the three types of existence of men: under the law of nature, under the law of Scripture, and under the law of grace*" (emphasis mine).[60] The italicized portion at the end of this passage communicates Olivi's understanding of *concordia*, that the Hebrew Bible prefigures not only the events and persons of the New Testament, but events throughout the whole of salvation history, here divided into yet another pattern of three.

Olivi's discussion of human existence may not, at first blush, seem entirely novel. It looks something like the usual binary between law and grace, but, as Burr rightly shows, it emerges from Jewish exegesis and had been employed by Christian authors prior to Olivi as a a typical means of "citing Judaism against itself." Olivi himself likely was drawn to the idea because the threefold structure is suggestive of its compatibility with Joachite apocalyptic thought, a point that is underscored by his employment of the same phraseology in his commentary on the Apocalypse.[61] Olivi, like Joachim before him, conceived of sacral history as unfolding in three ages—those of the Father, Son, and Holy Spirit. The schematic rendering of a threefold history by Jewish sages was simply more evidence that Joachim and Olivi were correct.

Olivi likely believed himself and his fellow Christians to be on the cusp of a transition to the third and final epoch. Such a transition would not obviate prior biblical writings, nor require additional scripture, but would include a deeper and more spiritual understanding of the mysteries encoded within the Bible. Hence, each era was aligned with a person of the Trinity and a moment of scriptural awareness.[62] It is likely that penetrating deeper into the biblical texts was precisely Olivi's point in writing. Hence, when Olivi writes on Genesis (or any other piece of scripture) he is never writing just about Genesis. Genesis is always in conversation with other scriptures, as well as fundamental theological truths, such as the nature of the Trinity, and salvation history.

Probably the best example of this comes early in the commentary, when Olivi discusses the six days of creation (and final day of rest) as a historical model. The six days of creation (and the day of rest) had been a model of salvation history since the patristic era, so Olivi is hardly breaking new ground, but he builds a more complex picture than seven phases of history corresponding with the seven days. For example, Olivi argues that the pattern of seven asserts itself in both Testaments. In the Old, history occurs in seven distinct phases, which Peter calls both a *tempus* and an *opus*: 1) the ornamentation of nature (that is, creation), 2) the events leading up to and including the Flood, 3) the covenant with Abraham, 4) the establishment of the Law, 5) the glory of the kings of Israel, 6) the time of the prophets, and 7) a time of quiet in which the Temple was rebuilt.

Likewise the seven days set a pattern for the *tempora* or *opera* of the New Testament and the period after it. Since not all this era is yet consigned to history, Peter uses the events of the first seven days to mystically (*mystice*) explain the unfolding of salvation history both past and future. On the first day, God created light; hence, the first *opus* or *tempus* of the New Testament is the sowing of grace, that is, the time of Christ and the apostles. The separation of the waters on the second day provides a metaphor for the "baptism" of the blood of the martyrs. On the third day, dry land appeared, corresponding to the early Church from Pope Sylvester through the general councils of late antiquity. Here the close visual and visceral connection to the days of creation and eras of breaks down, as the fourth and fifth ages belong to Justinian and Charlemagne respectively. The sixth *opus* is somewhat vague from a temporal standpoint, as it is the era in which Olivi positions himself, a time of deepening faith and the emergence of a Christ-like way of life. In regard to the latter, Olivi probably had in mind his own Order, but likely also other friars and monastic orders. He sees this period as having begun with the growth of formal theology as a discipline. Finally, the day of rest is a future event, the opening of the seventh seal as described in the Apocalypse.[63] While one might expect similar and very spiritual readings of Genesis to follow, instead we find that Olivi changes course quite quickly, and focuses almost exclusively on natural philosophy for the rest of his discussion of the Hexaemeron. The commentary on the six days alone is quite long, and summarizing its contents is difficult, as Olivi tackles a range of subjects. Instead, a few examples will serve to show how Olivi used his Genesis commentary as a means of explicating natural philosophy and truths regarding the natural world.

At times it seems that Olivi is using Genesis as an excuse to tackle ad hoc philosophical questions. One such example comes from the second of his two chapters on the six days of creation. Taking up Genesis 1:3 (*And let there be light*), Olivi jumps into a discussion of optics, specifically whether rays of light retract or not. "Some people" says Olivi, not naming his philosophical adversary whether real or imagined, "think that light does not move from its place—that after the space of a usual day when it has illuminated the world it retracts its rays and thus it becomes night in the whole world."[64] Olivi naturally thinks this is impossible, and argues that light rays are *accidents* of luminous bodies that do not retract themselves. Olivi argues that such an action would require a miracle of God, but he doubts this has ever occurred. Light emanates from its source, argues Olivi, and the kind of mutation required to make light retract itself would be unnatural. Olivi later connects this argument to the kind of light emanated by saintly bodies and by Christ, but they are examples of his theoretical claims rather than their aim.

In Olivi's desire to connect the seven days to salvation history, the Trinity, and other Christian truths, he never considers the notion that the six days were metaphorical nor only meant to be understood allegorically or spiritually. Olivi understood creation literally, as a six/seven day event in which God formally created the universe. It is this literal view that informs the naturalist aspect of his commentary, and also prompts him to take on the Augustinian reading of Genesis as a spiritual text rather than as an accurate account of the physical nature of the world. Olivi takes on Augustine early in his commentary, again when discussing Genesis 1:3. Here Olivi argues that the light of Genesis 1:3 refers to actual light and the division of the days rather than the enlightenment of angelic or human nature. Augustine was deeply interested in drawing out the spiritual dimensions and lessons of Genesis, a general sentiment to which Olivi was quite friendly, but did so at the expense of what Olivi considered to be a representation of the physical composition of the cosmos. This allows Olivi to agree with some of Augustine's more specific assertions, such as the fact that the Hexaemeron suggests the Trinity, and pay lip service to Augustine's insight. Both authors agree that Genesis is a meditation on the relationship between the divine and the human, but disagreement on the details overshadows Peter's claims of broad thematic agreements.[65] Augustine for his part sees Trinitarian unity expressed in the first two verses of Genesis. The pericope "In the beginning God made heaven and earth" contains two persons of the Trinity—the Father (God) and the Son or Word (the beginning).[66] The third person of the Trinity, the Holy Spirit, is found in Gen 1:2, "And the Spirit of God was being borne over the water."[67] This is, to Augustine, a "complete reference (*conpletam commemorationem*)" to the Trinity, already established before the work of creation, which is signified by *lux fiat* in Gen 1:3.[68] At issue for Augustine is the coeternal nature of the persons of the Trinity, a much debated point of theology in the early Church. Augustine therefore considers that the six-day course of creation was nothing more than an allegory and that creation occurred all at once. Olivi, for his part, sees the Trinity as being revealed over time. This is not to say that he denies the coeternality of the Trinity (he does not). This was, in Olivi's era, assumed to be true. For Olivi, the Trinity always existed, but scripture reveals the nature of the three persons gradually over the six-day work of creation, as part of a series of patterns of threes.[69] Temporality thus becomes the crux of Olivi's dispute with Augustine.

What bothers Olivi is that Augustine's contemplation of the text—that is, a broadly allegorical reading—does not meet Olivi's understanding of a literal commentary.[70] Olivi is so put off by Augustine's reading of Genesis 1:3 that he cannot help but blurt out that Augustine's reading is just flat out untrue according to any reasonable interpretation of the text. Taking an implied shot at the title of Augustine's work, *The Literal Meaning of Genesis*, Olivi continues, "It certainly isn't literal, since the whole of scripture prior to the first day up through the seventh clearly points out that it is talking about real days and nights." To be clear, Olivi stresses his agreement with Augustine that Genesis has a spiritual message, as his Joachite approach would suggest, and he often validates mystical readings of Genesis and relies on Augustine elsewhere.[71] Yet, Olivi makes a clear break with exegesis that, in his mind, appears to twist the words of scripture to mean something that a plain reading would not suggest, that is, Augustine's claim that everything was created in an instant and that the subsequent description of creation existed only to assert certain theological truths.

Where earlier generations of exegetes had passed over literal readings of the scriptures as a kind of preamble to the meat of exegesis, the drawing out of the moral and spiritual meanings of a text, Olivi was keen to suggest the value of the literal. Some of his concern likely stemmed from his attachment to Joachism. For one, the Joachite principle of *concordia* relied heavily on the chronology or chronologies of the Bible. Seeing harmony between the texts relies not only on a typology of persons, but also on typologies of events—

chronological and numerical patterns that emerge over and over within the text and, hence, within history. Augustine's compression of creation to a single moment of "God created the heavens and the earth" is antithetical to the parallelism of Joachite readings.

Olivi was also concerned with demonstrating the philosophical order of creation, asking whether or not the sequence of the six-day creation was a mystery or whether there was some other reason behind it. While admitting that there were many mysteries, he first put forth a rational and philosophical justification for how creation unfolds. The first three days, he argued, dealt with distinguishing undifferentiated confusion (via the *lux fiat*), and then fixing distinct bodies into place (dividing the waters and separating the land). This was also the period during which the terrestrial elements became distinguished from one another and heaven became incorruptible as a means of distinguishing it from the terrestrial world. The next three days dealt with "adornment," that is, the populating of the cosmos, with celestial bodies, birds and fish, and finally animals and human beings. Human beings came last, Olivi says, because they were the most important part of creation, and relied on everything created before.[72]

Olivi also weighed in on a number of additional issues that would have been critical to alchemists. For instance, Olivi frequently makes reference to the protean stuff of creation (*moles*). Olivi discusses it as an intermediate step (*intermedio modo*) in creation, but not necessarily as prime matter. Rather, Olivi seems to understand it as a confused mix of elements, all tangled up with one another.[73] They must have existed by the third day in some form, he argues, since Genesis does not say they were created (rather that they were separated or passed over). Yet, if they were fully formed, then Olivi believed Genesis would be talking only about accidents or particularities, rather than about creation. Instead, some kind of confused or indistinct matter existed.[74] It would be wonderful to know if Olivi thought this indistinct matter might be made by art, but Olivi does not oblige. What his text does show is that the Hexaemeron was a ripe source for speculation on questions of natural philosophy. While Olivi himself does not take up alchemical questions, it is clear that he is quite interested in in dissecting the composition of the world. His concentration on these questions, often to the exclusion of more spiritual readings, is very much in line with the kind of speculation recommended by Bonaventure.

The Franciscan exegete Nicholas of Lyra, writing a generation later, reflects Olivi's concern for the literal interpretation.[75] Nicholas's principal concern is to demonstrate the appropriateness and the order of the account of creation in Genesis. For instance, the mixtures of inanimate elements were made for the benefit of animate creatures, which likewise were made for the benefit of human beings. However majestic the act of creation, Nicholas wants his readers to see the obvious logic embedded in the creative act itself. Unsurprisingly, then, Nicholas relies heavily on Aristotelian terminology.[76] Following Averroës, he describes the work of creation as subdivided into discrete acts of creating new specific forms (*forma seu distinctio*), such as light. Light generally then, after its creation, is subdivided into specific lights, such as the various heavenly bodies. Nicholas describes this as the act of adornment and disposition in which specific qualities are attached to created forms.[77] It is important to note that the philosophical description of creation is Aristotelian rather than Platonic. The notion of *form* Aristotle uses in the description of light is not ideal. Nor is there a sense that the distinction of light (*lumen*) into lights (*luminaria*) diminishes their "lightness." Nicholas relies on other aspects of Aristotelian philosophy more specifically later in the text.

There are two points, however, that need to be emphasized in Nicholas's description of creation. First, not only is creation *described* in Aristotelian terms, but God *works* like a philosopher, carrying out creation according to a logic that is knowable to human beings. To Nicholas, this is the clear meaning of the text of Genesis, not an approximation. His literal reading of the six days is designed to tell his readers not just what to think about Genesis, but what really happened. And, in this case, God creates the universe according to an Aristotelian logic. By extension, then, Aristotelian accounts of matter, its mutability, and properties are also true. A second key point to emphasize is that this Aristotelian creation account is not a separate explanation from the sacral account of creation. Rather, the sacrality of creation is expressed within the frame of Aristotelian logic, and vice versa. For example, take the differentiation of the creation of form from that of adornment. According to Nicholas, this leaves us with three (rather than two) modes of creation expressed in the six days: creation, generally speaking, and the subdivision described above. General creation, he says, is described "in advance of each day (*ante omnem diem*)." Creation of form is described in the first three days, ornamentation in the following. Thus, what Nicholas has done is to overlay a pattern of two, three, and seven, neatly encapsulating the three most important numerical patterns of biblical exegesis. Nicholas was not a Joachite, but he interpreted the narrative of creation as an integrated Aristotelian account of the material world as well as an expression of fundamental Christian truths.[78]

Throughout the works surveyed here, there has been no mention of alchemy. Nevertheless, it is clear that Franciscan alchemists worked in an environment where discussion of natural philosophy was not only widespread, but curious, inventive, and appreciative. Not only did the inherently good created world lend itself to the revelation of the divine, but plumbing natural philosophy served religious, even devotional ends. In the sense that alchemy as theoretical knowledge explicated the building blocks of the natural universe, there is nothing about theoretical alchemy that is inherently at odds with the kind of discussions of nature seen here.[79] As a *scientia* of the physical world, alchemical concepts and ideas are wholly consistent with the kind of trends being discussed in the various natural theologies of Bonaventure, Olivi, and Nicholas. Beginning with their founder, Franciscans as a group revered the natural world. Moreover, scriptural accounts of creation could be fully in harmony with what natural philosophy could say about the cosmos. Pagan philosophers might have wandered into error due to natural philosophy, but that was not the fault of natural philosophy exclusively. Instead, what these accounts of the created world suggest is that pagan natural philosophers wandered into error because they were not armed with Genesis. The fact that theology faculties and clerical authorities condemned Aristotelian natural philosophy repeatedly in the thirteenth century should not obscure the popular notion that harmonious accord between philosophy and scripture was possible. The condemnations of Aristotle's *libri naturales* were, after all, aimed principally at wrong conclusions (either real or imagined), rather than expressing a disdain for knowledge of the natural world.[80]

If, as seems to be the case, Franciscans were reading Genesis as a critical text in natural philosophy, it seems unlikely that alchemical speculation could have been free from religious overtones. While other scholars have argued that we should take seriously the introduction of religious language during and after alchemy's religious turn—an idea with which I wholly agree—we should push somewhat further.[81] The religious turn was not solely or simply an irruption in the alchemical tradition. When viewed through the lens of hexaemeral commentaries and Franciscan concern for "the book of world," the religious turn emerges as a logical consequence of broader, even normative, intellectual trends. The explicative function of religious metaphor laid out in the introduction tells only part of the story. Alchemy in the fourteenth century was also catching up with, and innovating within, a discourse on the created universe where the divine was written into the physical fabric of the world.

CHAPTER 2

Three Elixirs

This chapter is about genealogies. What I hope to show is the importance of the intellectual components of the elixir—the alchemy of the making of alchemy, if you will forgive the expression. My chief concern is with the religious elements that make up the genealogies of the three elixirs under discussion, but these intellectual components mix and aggregate with natural philosophic components as well. The descriptions of the elixirs involve a number of overlapping discourses, traditions, and very personal ideological commitments. While religion need not figure into any conception of the elixir, the nuances of theological or philosophical speculation have serious implications for how the various authors envisioned their alchemies. Roger Bacon relies on the Christian model of the perfect, resurrected body. Vitalis of Furno draws on the tradition of the theriac, but studiously avoids mention of alchemy, and John of Rupescissa ruminates on the physical nature of heaven. These small differences are important. One cannot disentangle the "science" from the nonscientific assumptions and traditions that inspire the creation of these substances without fundamentally misinterpreting them. For while it is true that every elixir is a cure-all, each elixir is not just a cure-all. Genealogies clarify what generalizations obfuscate.

The pursuit of the elixir was one of two main tracks pursued by medieval alchemists, second in popularity to the transmutation of metals. Chrysopoeia and argyropoeia (transmutation of base metals into gold or silver, respectively) were the subjects of most medieval alchemical texts. The term elixir is derived from the Arabic *al-iksīr,* itself a translation of Greek *xērion,* a term used to describe the agent that transforms one substance into another.[1] Initially in Arabic alchemy the term "elixir" had nothing to do with healing human bodies, but rather with perfecting metals. Metals were understood to exist in a hierarchy, with gold at the top, followed by silver.[2] Medieval people generally differentiated metals phenotypically, that is, based on their appearance and observable characteristics, but medieval scholars also understood metals to have different elemental compositions related to primary qualities. Aristotle described four primary qualities (hot, dry, wet, cold) from whose interactions were made the four elements (fire, earth, air, water), which in turn, according to philosophers and physicians of antiquity, combined to make up the four humors. Hence, an *al-iksīr* perfected a metal by purifying it of its base qualities, but also by changing its primary qualities.[3] For instance, tin is primarily dry, but gold is hot and silver is cold. An elixir can change those qualities. A medical elixir worked according to the same logic. It both purified the body and balanced or redistributed the body's humoral qualities into a better arrangement.

While Western, Christian alchemists drew on the general idea of the elixir, they elaborated on its powers and production in ways not anticipated by their Islamic forebears. They seized particularly on the power of the elixir to heal and augment the human body. The justification for the potency of the elixir came from many sources, and the three elixirs presented in this chapter have very different alchemical and medicinal genealogies. Indeed, "elixir" was not a standard term for alchemical medicinals in the medieval period. Rather, authors named compounds "inestimable glory," "burning water," and "quintessence." Still, the logic of purification and balancing or transforming humors was characteristic of these substances. Therefore, "elixir," as I am using it, is simply a taxonomic term to denote an alchemical compound that has salubrious effects on the human body. That said, the terms used by the authors are important, and will be discussed in due course.

Knowledge about the elixir, and alchemy in general, emerged sporadically throughout the twelfth century. The first complete alchemical text to be translated from Arabic into Latin was likely the *Morienus*, completed by Robert of Ketton in 1144.[4] Its preface states that it includes knowledge on alchemy "our Latinity does not yet hardly comprehend." Evidence of the unfamiliarity of Latin authors with this new science comes from the occasional use of Western vernaculars as "intermediary" languages. Arabic, Hebrew, and Greek texts often found their way into local dialect prior to being translated into Latin.[5] Rather than diffusing alchemy into vernacular culture, these intermediary translations were a means of coping with what must have seemed entirely new concepts.

As important as the translation of Arabic alchemical texts was the translation of the Aristotelian corpus, by which I mean of course not only the works of Aristotle, but works ascribed to him and commentaries on all these. One of the key texts was the *Meteorologica*. Gerard of Cremona had translated its first three books early in the twelfth century, and Henricus Aristippus the fourth in 1156. But the work did not take its full shape in the Latin West until after 1200, when Alfred of Sarashel added to the manuscript a copy of Avicenna's *De congelatione et conglutinatione lapidum* (*On the congealing and joining of stones*), a part of his *Book of Remedy*.[6] What was probably the first original Latin treatise on alchemy, the 1257 *Book on the Secrets of Alchemy* of Constantine of Pisa, relied heavily on the fourth book of the *Meteorologica*, which included Aristotle's theory of metals.[7] Likewise Paul of Taranto, a Franciscan lector, authored early alchemical treatises that lean on Aristotle and his commenters.[8] Paul (under his own name as well as that of "Geber") also transmitted a theory on the generation of metals that linked them to the heavens. Contemporary astrological theory held that terrestrial objects were governed by heavenly bodies for which they had an affinity. For metals, the sun was related to gold, the moon to silver, and, predictably, Mercury to mercury. *The Book of the Secrets of Alchemy* includes this idea, but also holds that the celestial bodies were the origin of these metals.[9] The idea was that heavenly rays penetrated the earth and congealed within it as metals, helping to explain the affinity between celestial and terrestrial.[10] The celestial-terrestrial relationship would prove fertile ground for future alchemical speculation.

Roger Bacon and Inestimable Glory

In 1268, Bacon completed his *Opus tertium* (*The Third Work*), as part of a hastily written collection of texts that responded to a 1266 command of Pope Clement IV. The *Opus tertium* was a recension of the *Opus minus* (*The Lesser Work*), itself a recension of the *Opus maius*, a hastily written but voluminous tome he had sent to the pope probably in 1267. In this third work, Bacon writes that, for fear of alchemical secrets falling into the wrong hands, he has consigned some of the most vital secrets to his aide, John, from whom the pope can have this knowledge transcribed.[11] He also says that in addition to what John can convey verbally, there are two other sections Bacon had written obscurely, one in code (*enigmatas*) and the other in philosophical language so that a reader would assume Bacon was discussing medicine or natural philosophy rather than alchemy.[12]

We should not take this to mean that Bacon is short on detail or secretly skeptical about alchemy. In the short period since he had written the *Opus maius* after promising the pope he had already completed it, Bacon seemed more sanguine about alchemy, a notion bolstered by his later writings.[13] In fact, he had taken a cautionary tone regarding alchemy in his *Opus maius*. Speaking about the medicinal use of alchemical gold, a substance Constantine of Pisa believed might be capable of halting the spread of leprosy, Bacon opined that the gold produced by alchemists was not of a particularly high grade.[14] Interestingly, however, he said the same thing about gold found in nature, allowing that certain scientific arts might allow for the production of an even higher grade of gold. No doubt one of the reasons Bacon felt the need to create the *Opus minus* and *Opus tertium* was his evolving notion of alchemy.

In the *Opus maius*, Bacon had yet to align the practice of alchemy with what he called *scientia experimentalis*, troublingly translated with the cognate *experimental science*. By the time he wrote the *Opus tertium*, however, Bacon had divided the practice of alchemy into what he called "speculative" and "operative" alchemy. Operative alchemy was practiced, but not (fully) theorized. Operative alchemists could be skilled practitioners, but they did not comprehend the primary goal (*finem principalem*) of alchemy.[15] Speculative alchemists, of whom Bacon said there were but a handful, understood the uses to which alchemy could be put in regard both to inanimate matter, such as metals, dyes, and tinctures, but also to animate or living matter. Living matter, chiefly the human body, was the province of the speculative alchemist.[16]

Living matter corresponded to inanimate matter, but was more complex.[17] Like other Aristotelians, Bacon understood the four basic elements as arising from the four primary qualities (cold, dry, hot, wet).[18] From the elements come the simple humors, which correspond to the standard humoral model: phlegm, choler, melancholy, and blood. The simple humors have "conjoined natures" (*conjungentes naturae*) of the elements, that is, one element and its particular qualities (cold and dry, hot and dry, cold and wet, and cold and dry) dominate.[19] These humors, however, are not the same as bodily humors. The bodily humors are themselves a combination of simple humors. Each is dominated by one of the simple humors. So, bodily phlegm is a secondary humor, made up of four primary or simple humors, with primary phlegm dominant. Bacon's system, then, makes animate bodies more complex expressions of humoral interaction than those found in inanimate bodies.[20] To make true blood, one had to purify bodily blood of the other simple humors mixed within it.

Given the relationship of humors to elements, Bacon considers alchemy one of the essential fields of knowledge for understanding living things, so long as it is married to *scientia experimentalis*.[21] Jeremiah Hackett has distilled Bacon's description of experimental science into three "prerogatives."[22] The first is the rigorous application of logic to observation and experience. The term "experience" is, in fact, closer to the meaning of *experimenta* than is the cognate "experiment." Bacon argues that observation of phenomena can teach us new truths and even refute truths supplied by reason alone, as long as we apply our critical faculties to what is being observed. The second aspect of experimental science is the creation and manufacture of tools, weapons, and medicines. For instance, in addition to the elixir, Bacon proposes using experimental science to create giant mirrors capable of incinerating enemies at a great distance.[23] Finally, Bacon believes that experimental science can supply a reliable source of divination for knowing things of the past, present, and future. This generally relates to the observation of astrological phenomena, but his writings on the elixir suggest that human beings may be able to gain such knowledge instantly.[24]

This is a good indicator that Bacon meant something rather different from experimental science by *scientia experimentalis*. In the *Opus tertium* he even calls certain branches of *scientia experimentalis* "magical (*magicus*)."[25] In one sense, this is atypical of Bacon's work, since he generally uses the term "magic" to describe deceits.[26] On the other hand, the breadth of sciences Bacon recommends includes a number of practices that are defined, certainly today and often in medieval culture, as magical. Divination and the utterance of "words of power" are just as much a part of Bacon's program as geometry. Thus, Bacon's use of the term "magic" tends to have ethical and religious overtones. The medieval approach to magic was to define it in opposition to religion. Modernity retains the tendency to define magic as part of a binary, but the modern definition tends to contrast it with science.[27] The problem here is that, well into the early modern period, the magic-science binary was not distinct, given the conviction that, as Michael Bailey has put it, "the natural world was conceived as a direct manifestation of the supernatural order."[28]

Bacon is quite aware of the problem of terminology. One of the reasons he advocates so strongly for *scientia experimentalis* is that too many of its practices have been discarded as magical or sorcerous, when they should belong to the natural philosopher.[29] Bacon argues that the application of *scientia experimentalis*, especially critical observation, can help sort out what is magical and what is not.[30] Bacon uses the example of a magnet to prove this point. People are easily fooled, he argues, when a fraudulent magician uses a magnet to attract iron. By uttering incantations or scrawling sigils on the ground, the magician makes people think that these activities move the iron, when it is in fact the natural property of the magnet.[31] On the other hand, Bacon does not reject magical words or symbols wholesale, griping at one point that women and demons have "abused characters and incantations written by the wise."[32] Therefore, *scientia experimentalis* enfolded a wide variety of practices, linking them through a particular methodology.

One linkage in particular bears special mention, namely, the relationship between astrology/astronomy and medicinal or speculative alchemy. As noted in the prior section, the connection between celestial bodies and terrestrial bodies (of any kind) was understood to be quite strong. Bacon himself was a fierce defender of the merits of astrology and reinforced the link between celestial bodies and human physiology.[33] The effect of celestial motion on the human changed from moment to moment, and intersected with a person's humoral make-up, what Bacon called complexion. The effect of both the stars and the humors (themselves partly determined by the position of the stars at one's birth) exerted significant force on human behavior. The notion that physiology was somewhat determinative in regard to behavior has a history dating back to Aristotle, but Christian theologians had resisted the idea of deterministic forces compelling behavior. One had to be free if sin was to be a meaningful category. Bacon is careful to maintain that complexion inclines people to action, but does not obviate free choice.[34] Rather, Bacon hoped to use astrology and alchemy to mitigate the negative effects of the humors and stars and intensify the positive.

Bacon took a moderately declensionist view of human bodies and their complexion. He agreed that biblical writings demonstrated that humans once had much longer lifespans, which had shortened gradually since the exile from Eden. At the same time, he did not think that this was a necessary condition of human existence. Sinfulness and immoderation had polluted humanity's ancestors, who passed on these defects to their children, where they were compounded over succeeding generations and shortened the human lifespan. Because this was an "accidental," not a necessary, condition of the world, human lifespans were subject to expansion as well.[35]

Bacon recounts a number of tales (hearsay for the most part) to prove the truth of this statement. (As much as Bacon championed observation and reason, he was also prone to rely on the authority of anyone worthy of belief.[36]) For instance he relates the tale of a peasant who found in a field a golden vessel full of liquid that bestowed on him sixty years of youth.[37] He also discusses the benefits of consuming dragon flesh, including both health and mental acuity.[38] These medicines might seem miraculous, but Bacon regards their effects as resulting from their composition: "That liquor which the rustic drank is thought to have approached an equality of elements far beyond ordinary food and drinks."[39]

This equilibrium is one of two theoretical keys to Bacon's elixir. As he puts it in the *Opus maius*, "If the elements should be prepared and purified in some mixture, so that there would be no action of one element on another, but so that they would be reduced to pure simplicity, the wisest have judged that they would make a perfect medicine."[40] Such perfect mixtures are supposedly inaccessible through nature, making the administration of medicine a fraught pursuit. One of the concerns of medieval pharmacy was to ensure that medicines did not further disrupt the bodily imbalance that caused the ailment in the first place, or reverse it in such a way as to cause a different ailment or death. Apart from physical injuries, medieval observers considered illnesses to arise out of a humoral imbalance. Too much blood, for instance, (a hot and wet substance) could produce certain kinds of fevers. Hence pharmaceutical documents engage in a great deal of discussion of the properties of various herbs, stones, foods, and liquids that might go into a medicine. What might be healthy for a feverish woman may not be good for a healthy man. Yet an elementally balanced compound, such as Bacon's elixir, had the potential to restore humoral equilibrium by purifying the body of the excess substance and conferring upon it what it lacked, regardless of a person's original complexion.

Perhaps the most powerful example Bacon employs to explain this theory is an example known to all his readers: Adam in the Garden. Adam, writes Bacon, was immortal only so long as he ate from the Tree of Life. Adam's body was, like all terrestrial bodies and compounds, made up of elements and humors that acted upon one another. It was subject to change, that is, to corruption, and therefore to dissolution. Because the humors acted upon one another, there could be both imbalance and waste. Hence Adam, like all people, needed to eat to rebalance his humors. Regular food, as noted earlier, has particular humoral qualities as well. While a thoughtful diet (that is, one based on food that replenishes lost humors and mitigates excess humors) is vital for health, it never perfectly balanced one's complexion. The fruit of the Tree of Life, however, had "elements approaching equality," and could have kept Adam alive indefinitely.[41] Just as we have seen in Chapter 1, Bacon is reading Genesis as natural philosophy. Bacon agreed that the interpretation of scripture relied on natural philosophy, but here we get a clear example of Christian doctrine driving alchemical theory. Had the Bible been a text that merited only spiritual reflection, it is hard to conceive of Bacon developing the elixir along the lines he did.

Adam's body was, of course, superior to postlapsarian ones (Bacon predictably makes no mention of Eve's body), but in one sense all bodies are the same. Every human body is naturally immortal (*naturalis immortalis*), meaning it can exist indefinitely provided it has a balanced complexion.[42] Bacon is again following scriptural warrant, specifically Paul's First Epistle to the Corinthians, where the apostle describes the resurrected dead as uncorrupted (*incorrupti*) and immortal. (Vulg. 1 Cor 15: 50–54) The fact that the resurrected dead must have perfect, uncorrupted, and incorruptible bodies is the foundation of Bacon's notion that human body is capable of being perfected.[43] What Bacon adds to scripture is a philosophical explanation for this idea. The resurrected dead, according to Bacon, are endowed with an equality of elements, which allows them to exist physically forever.[44]

This is not wild extrapolation or speculation on Bacon's part. Augustine ruminated on resurrected bodies, and Peter Lombard included a discussion of the physiology of resurrected bodies in his *Sentences*, the principal theological textbook of the scholastic era.[45] A number of Franciscan masters, including Bonaventure, composed (or dictated) commentaries on the Lombard's sentences, and any student in theology in the thirteenth century could expect to listen to masters' lectures on individual questions for years. Bacon's use of the resurrected body as a model for the effects of the elixir was significant, not only because of the impact of religion on alchemical theory, but also because the discourse was familiar to his confreres and the intellectual elite. Bacon's theory of the elixir was not particularly influential, but that does not mean it was misunderstood or obscure. Moreover, Bacon was quite careful to maintain that these perfect bodies would die when God wanted them to (though not due to illness or old age).[46] This accorded with the kind of perfect (resurrected) bodies found in some contemporary hagiographies.[47]

The resurrected, and therefore complexionally balanced body was characterized by four specific qualities, called dowries (*dotes*), so labeled by William of Auxerre and discussed in the *Sentences*.[48] The dowries are *claritas*, *agilitas*, *subtilitas*, and *impassibilitas*. While Bacon does not refer to these per se, it is clear from his description of the elixir's effects that the dowries are on his mind. While the dowries are often translated by their English cognates, it is important to remember that scholastics considered each of them to encompass a number of different qualities.[49] The gifts are not just infusions. They also speak to the removal of defects. Elixirs are substances that not only purify through the removal of unwanted qualities, but also confer what is missing. Thus Bacon's belief that an elixir can act to create a perfect body is wholly consistent with both scholastic dialogue on perfect bodies and alchemical theory.

Impassibilitas has a double meaning. On the one hand it is related to the Greek *apatheia*, an inability to suffer as well as a freedom from base passions. Thomas Aquinas refers to it as *quies*, freedom from the passions, and Bonaventure calls it a "perfect disposition."[50] Physiologically speaking it meant imperviousness to corruption. The composition of elements within an impassible body could not be changed. Bacon provides an example of this idea by discussing the not so fortunate resurrected bodies in Hell. Their flesh could burn eternally while never being consumed by the flames.[51] *Claritas* describes the luminous quality of perfect bodies—a well-attested aspect of the resurrected dead and of saintly bodies.[52] *Subtilitas* speaks to the refinement of the body's particles to the point that these fine particles can pass intact through other bodies; the resurrected can walk through walls. *Agilitas* is the ability to move one's body according to the wishes of the soul without limitations, to float or levitate. Bacon does not delve much into these latter three dowries, though there are some implications of *claritas* when he discusses how one concocts the elixir.

Thus far, I have discussed Bacon's elixir as if it were a simple alchemical compound capable of producing a resurrection body, but this is not a full explanation. Creating the elixir did require alchemical skill. Blood mixed with mercury, a noble metal (usually gold), and a base metal were a starting point, and the alchemist was expected to use blood to extract primary humors necessary to the preparation of the compound.[53] Artisanal skill was not sufficient to bring the elixir to completion, however. Otherwise the operative alchemist would be able to make such a medicine. The elixir also required celestial ingredients, which would be accessible only to those trained in astrology and *scientia experimentalis*.

In his introduction to his translation of pseudo-Aristotle's *Secret of Secrets*, a popular text that was highly influential on Bacon's later work, Bacon advocates that physicians expose their medicines to the heavens. When favorable constellations or celestial arrangements are present, medicines should be taken out of doors. Otherwise they should be kept indoors.[54] The same regimen goes for patients. The logic is plain enough. If planets, stars, and constellations govern, or at least influence, various terrestrial counterparts, then the physician should take advantage of this. Astrological medicine had a long tradition in the West, but Bacon, as he did with other inherited ideas, was redeploying these concepts for his own ends rather than simply repeating them.[55] Bacon's elixir is completed, then, when infused with the beneficial properties of the heavens. Chief among these properties would seem to be incorruptibility, but Bacon also has another benefit in mind—to use celestial rays on people to "make better their complexions so that they are inclined to good and useful things both for themselves and for others, in wisdom just as in morals."[56] It is important to bear in mind here the link between humors, stars, and behavior. Though there were many variations of humoral theory, they all posited a unified individual, where body and mind were inextricably linked. The humors circulated constantly through the body, impacting one's personality and predilections. Likewise, celestial movements influenced humors, both at birth and throughout the course of one's life. The duality of the soul and body in some Christian literature often obscures this fact, but no one in the Middle Ages seriously disputed the humoral impact on the mind. Quite the opposite, in fact, as humoral theory was invoked to prove the natural fitness of men to intellectual and spiritual work. Likewise, as much as churchmen argued against astrological determinism, the influence of the heavens was not in serious doubt.

The elixir's effect on the corrupted human proceeds from body to mind and, finally, to morals. Bacon asserts, "if this greatest thing can be done [the production of the elixir], it is evident that all other things are possible, namely that man might reach great foresight and perfect wisdom that he may know how to rule himself and others, with the help of the grace of God."[57] This is a significant passage. It is clear from this that Bacon is endorsing no mere medicine for bodily ailments, but something almost salvific. Bacon is adapting—Christianizing, even—one of the claims made in the *Secret of*

Secrets. This text presents itself as a long advisory epistle from Aristotle to his pupil, Alexander the Great. Among the many recommendations is a method of pacifying the inhabitants of a particular conquered region by changing their complexions and that of their environment.[58] (Regions had complexions just as much as did discrete objects, and these complexions in turn affected the complexions of the inhabitants.[59]) Pacification is, according to Bacon's idiosyncratic understanding of the text, accomplished by "fascinating" the people of the region, "so that they were not able to help themselves," and by changing the complexion of the region to effect a change in the morals of the inhabitants.[60] Given the scant detail for how to accomplish this, Bacon is left to his own devices to come to a solution. Given his understanding of the effects of alchemy and astrology on the human body, it is natural that the elixir be used. In the *Book of Six Sciences*, he declares that changing people "in body and soul (*anima*), so that in them is furnished the natural goodness of longevity, character, foresight, and wisdom" is the "secret of secrets and the ultimate secret," that is, the true aim of alchemy.[61]

Some attention needs to be given here, however, to Bacon's use of the term *anima* or soul. Given that the soul is viewed as a free agent regardless of one's humoral or astrological condition, what is it that Bacon means here? It could be answered that Bacon is simply being sloppy in his vocabulary, engaging in a rhetorical flourish. Such an idea is hardly out of place. As we have noted, Bacon uses the term "magic" rather carelessly, and there are other examples of such imprecision. On the other hand, Bacon might very well mean to effect a change in one's soul. This requires some explanation. Perfect repair of the holistic body means perfect repair of the intellect, such that a human being enhanced by the elixir would possess "great foresight and perfect wisdom (*magnam prudentiam et sapientam*) . . . with the help of the grace of God."[62] Cognition was a power of the soul, and disembodied as well as embodied souls were understood to have different kinds of cognition.[63] Therefore, anything that improves the intellect must interact with the soul in some way.

Moreover, Bacon's claim that this change is dependent on God's grace is also significant. Bacon had earlier written in the *Opus maius* that *scientia experimentalis* does not provide complete knowledge, even of corporeal things, save through divine illumination.[64] Bacon ascribes this kind of illumination to the patriarchs and prophets as well as to many (unnamed) persons since the time of Christ—saints in all likelihood. Though Bacon credits (pseudo-) Ptolemy, the aphorism of the *Centiloquium* on which he relies does not develop this idea.[65] Given Bacon's preference for using the resurrected body as the model for the elixir, he probably has in mind something like what Bonaventure called the "perfect knowledge" of the resurrected dead.[66] For Bacon this entails at least hidden things of the past, present, and future, but the elixir also excites the rational soul to comprehend all sciences.[67]

Bacon, like all his contemporaries, agreed that spiritual and moral knowledge exceeded all other forms of knowledge, and in fact emerged from other kinds of knowledge. Moral science (theology) is the ultimate aim of *scientia experimentalis*. Just as Bonaventure advocated that comprehending the world led one closer to the divine, Bacon argues that terrestrial knowledge is the first rung on the ladder to moral and interior knowledge.[68] This included not only knowledge of virtue, but also possession of the gifts of the Holy Spirit and even rapture. Therefore, the elixir grants not only foresight and intellectual acumen, but also the most important kind of knowledge—knowledge of God. Bearing in mind Bacon's context, knowledge of God means knowledge of the Christian God and the truths of Christianity. Therefore, the use of the elixir produces not just a perfect body, but a perfect Christian as well.

Bacon's hope for the elixir goes beyond producing perfect Christian bodies one by one. Rather, he argues for making changes on the grander scale described in the *Secret of Secrets*. In his *Book of Six Sciences*, Bacon argues for something like a cumulative effect of the elixir. Once the experimenter has managed to balance his own complexion, his intellect will have advanced to the point that he will be able to devise a means of collecting, multiplying, and transmitting the same celestial rays that infuse the elixir. Bacon envisions the construction of giant mirrors that will send these celestial emanations over great distances, thereby changing complexions of regions and people on a massive scale.[69]

Bacon's goal is the alchemical-astrological transmission of Christianity or, at the very least, making people and places hospitable to the Christian message. While we will explore this idea in relation to Bacon's apocalyptic thought in the next chapter, it is worth noting that Bacon's understanding of the elixir goes well past its alchemical roots. In many respects, however, Bacon's elixir is thoroughly Franciscan. Bacon's investigation and attention to the natural world and the gifts of God inherent in it are very much connected to the kind of philosophical discussions found in Chapter 1. Where Bacon has traveled somewhat farther than his brothers, however, is in reifying the inherent goodness in God's creation. By anticipating a substance than can purify body, mind, and soul, Bacon is taking a logical, if perhaps unanticipated, step in the celebration of the created world. Where Bonaventure and Olivi scoured natural philosophy for indications of God's presence, Bacon is willing to literally scour the world for material evidence of God's plan. This is a leap his brethren, by and large, were unwilling to follow, but his discussion of alchemy nevertheless belongs to the wider Franciscan discussions of the created world.

Alchemical Medicine: The *Pro Conservanda Sanitatis* Ascribed to Vitalis of Furno

In the generation following Roger Bacon, Vitalis of Furno (d. 1327), a Franciscan Cardinal, wrote (or plagiarized) philosophical texts, exegetical works, *consilia*, and possibly a medical encyclopedia. Within this encyclopedia is a description of the *aqua ardens*, a compound that resembles the elixir. One generation later, the friar John of Rupescissa would write an entire treatise on the *aqua ardens*, which alone makes Vitalis's description worth examination. Just as important, however, is the fact that the *aqua ardens* presented in *Pro conservanda sanitate* (*On the Preservation of Health*) is a naturalistic compound. In regard to the influence of religion, this is perhaps the only place where we might see it as constraining, given the fact that—if Vitalis was the author—he lived through (and participated in) a turbulent period for the Franciscan Order and the beginning of a backlash against alchemical endeavors.

Vitalis's ascent through the academy to the upper echelons of the ecclesial hierarchy has led to some cynicism regarding his academic endeavors. Stephen Marrone, having examined Vitalis's theological-philosophical works, calls the Cardinal "an inveterate plagiarizer," a "careerist," and "never engaged deeply enough in ideas to be bothered with originality." Marrone finds Vitalis's philosophical and theological corpus almost entirely dependent on that of Henry of Ghent, a neo-Augustinian, whose work Vitalis not only followed, but relentlessly copied into his own treatises. Again, Marrone: "he incorporated stretches of other authors' writings into his own work so frequently and on so vast a scale as to make him exceptional even at a time when such unattributed copying was established practice."[70] Thus, when we refer to Vitalis as an author, we shall do so in the most liberal of terms.

Ascribed to Vitalis is a widely varied amalgam of writings, many of which cannot be attributed with full confidence, even less for the *Pro conservanda*. The primary argument for attributing the text to the Cardinal is that it bears his name, but that is hardly a reliable barometer of authorship. Other evidence is circumstantial. Just before the turn of the fourteenth century, Vitalis taught at the University of Montpellier, where he quite possibly had engaged in medical studies. Montpellier, at that time, had eclipsed Salerno as the most eminent school of medicine in the Latin West and was a center of discussion on the compound medicine called *theriac,* which the *aqua ardens* in *Pro conservanda* resembles. Another piece of circumstantial evidence comes from Vitalis's inveterate plagiarism. *Pro conservanda* itself borrows liberally from at least one near-contemporary encyclopedia, also penned by a Franciscan author.[71] These facts offer at best a rationale for *not* rejecting Vitalis as author out of hand, though they hardly offer much in the way of support for a claim of Vitalis's authorship. Barring a manuscript discovery, or other external evidence coming to light, we must approach the text cautiously. Given the certain Franciscan provenance of much of the evidence to be discussed, however, it merits inclusion in our discussion, since Vitalis's putative corpus serves as a normative counterpoint to the more idiosyncratic works of Roger Bacon and John of Rupescissa. The texts are unoriginal and were unlikely to cause a stir, as was likely the point.

The character of *Pro conservanda* is encyclopedic, though it is far from exhaustive. The subject matter is arranged alphabetically, with no systematic theory or apologetic intent immediately apparent. Constantine the African and Aristotle are among the more common authorities cited.[72] As such, it is eclectic, but not odd. Much of what one might expect to find in a learned medieval medical tome is here, though it also bears much resemblance to the encyclopedic work on which much of its material depends.[73] Given the compilatory nature of *Pro conservanda*, internal evidence points to no specific author.[74] One can make certain assumptions, however. For one, the author (or compiler) was educated. The text is composed in Latin, and includes Aristotelian themes.[75] Second, the author had access to medical knowledge that was theoretical in nature. This latter point in particular suggests a university-educated author, possibly one from Montpellier, where Vitalis spent some of his life.[76] It also appears that the author was either well read or well traveled, as the text occasionally points out regional variations, for instance, the use of fennel by Italians to promote gastrointestinal health, suggesting a non-Italian source.[77]

One problem with identifying the author or compiler of this text is, as mentioned already, the fact that it leans very heavily on the thirteenth-century encyclopedia, *On the Properties of Things*, composed by Bartholomaeus Anglicus. In nearly all the sections of *Pro conservanda* I examined, I discovered that a substantial portion of the text relied on the encyclopedia.[78] Sections of the encyclopedia often were lifted verbatim, though the author of *Pro conservanda* frequently rearranged them and put them under different headings. Given Vitalis's penchant for lifting whole sections from the work of the Parisian master Henry of Ghent without attribution, a cynical observer might see this as evidence of Vitalis's authorial role. Whether or not Vitalis himself composed the text, the Franciscan Bartholomaeus must be considered one of its authors.

Pro conservanda is both a theoretical and a practical text, though it leans more toward the latter, since explanations of why certain cures are efficacious are not always present nor are discussions of etiology (though in most cases a short explanation is supplied). For instance, the text recommends that if someone is mad (*phreneticus*) or suffers in the head or one of the limbs, the recommended course of treatment is to tightly bind the person. In this way, the pain of the binding will diminish the suffering elsewhere. A sound, if not particularly inviting, principle, but it is typical of the type of reasoning present in the text. It goes on to explain that the diminishment of the pain owes to the flowing of the humors away from the initial site of the pain and into the constricted limbs.[79]

Descriptions, etiologies, and remedies emphasize the natural, even when dealing with the super- (or preter-)natural. Take, for instance, the heading on *Sorcerers* (*Malefici*), which is brief enough to quote in full:

> Sorcerers enjoy the friendship of Hoopoes, which are a bird dwelling especially in sepulchers, tombs, or in dung. And they are especially unclean, (Lev 11:13–19) with feathers sticking out of their helmeted head. And whosoever anoints himself with its blood, he will see (*videbit*) demons strangling him. And its heart is useful to sorcerers, for they use it in their magic. The bat, an animal similar to a rodent, like a bird with unfeathered wings, flies in twilight with exceptionally quick flight. It dwells in the fissures of walls, is of a cold nature, and drinks the oil in stones. Wherefore the blood of the bat, when smeared on the eyelids stops hair growth, since it blocks the pores

> with its frigidity, and said pores cannot re-grow hair. See also *frog*, *fox*.[80]

In an entry on sorcerers, we learn very little about sorcery. Instead we learn that the hoopoe and bat are unclean and that their blood has deleterious properties, owing, in the case of the bat, at least, to its humoral composition. The demonic strangulation is likely an apparition, though no doubt terrifying, and there is a general reference to use of magic without any specifics. This is as close to a description of supernatural or infernal agency as the author of *Pro conservanda* comes, which rather starkly contrasts with the demonic etiology of illness we will see in the work of John of Rupescissa.[81] The naturalism of *Pro conservanda* is one of its defining features, if not a particularly novel one, as is the sometimes off-topic discussion which owes to the eclectic nature of the text's assembly.

Alchemy has no specific heading of its own in the text. There are, however, three areas worth exploring to understand how alchemy is treated: the entries on gold and silver, planets, and waters. Since the latter entry contains the only overt information on medicinal alchemy, I will dispense with the sections on gold and silver and planets first.

Though a significant portion on the salutary effects of precious metals for human health has been lifted from *Properties*, the initial discussion of gold and silver has nothing to do with healing at all, but with forgery.[82] The description of the forgery techniques is not particularly compelling, but dips into alchemical theory. Forgers place their gold and silver in cold and damp environments. Since these metals attract cold and damp humors, this causes them to weigh more until they dry out.[83] The prospect of alchemical forgery was a real fear by the early fourteenth century, eventually culminating in a papal bull, *De crimine falsi* (*Spondent*), condemning alchemists, by John XXII in 1317.[84] The language of the bull says more about the nature of alchemists than it does about alchemy itself. Alchemists, the bull, maintains, are charlatans, whose forgeries may look like gold or silver, but are not.[85] The bull denies the power of alchemists, but not alchemy per se. One of the reasons alchemists deceive others is that they are ignorant. The implication is that transmutation may be possible, but that is not a capability of most alchemists. The bull takes pains to demonstrate that its intent is to protect the common folk, though a more likely explanation is the fear of the papal treasury being filled with counterfeit coinage. [86] Forgery has two components, the use of metals (*metallis*) for the coinage and the actual forgery of the stamps (*characteres*) that embossed them.[87] It is important to note that the condemnation is limited to forgery of currency, and is not a broadside against alchemical theory or practice. While at one point the bull states that metallic transmutations are against the precepts of law, the papal jurist Oldrado da Ponte, active during John's papacy, dismissed this sort of argument.[88] He stated that such transmutations occurred in nature and that alchemy could not be condemned based on violation of such principles, further weakening the bull as any kind of broad attack on alchemy.[89]

One other aspect of the bull merits note. The bull takes special note of clerks who are engaged in this practice—whether alchemy or just forgery it is not clear, though I would favor the latter. [90] These unlucky souls, in addition to the normal penalties, were to be deprived of their benefices and prevented from holding new ones.[91] This speaks possibly to official concern of ongoing clerical involvement in alchemy, but more certainly to a fear that the tithes flowing to St. Peter's might be forged currency.

Two additional bulls promulgated by John XXII shed some additional light on the curia's view of alchemy around the time Vitalis was active. The bull *Super illius specula* (1326/7), which condemns magical practice and sorcery, does not include the alchemical arts. Nor are alchemists necessarily among the "great many people, Christian in name only, who . . . enter into a treaty with death and make a pact with hell."[92] Clearly those making "images, rings, mirrors, phials or other things of magic to bind demons" are not the same as those making counterfeit currency, nor does alchemical forgery apparently require demonic aid.[93] Rather closer to the text of *De crimine falsi* is that of *Prodiens*, another bull aimed at counterfeiters, though technically limited to forgers of currency and its importers in the realms of France and Navarre.[94] Alchemy is not mentioned in this bull, so clearly no one believed alchemical practice was required to forge currency. Rather, what this demonstrates is that in this era the primary danger of alchemy, so far as the curia was concerned, was the ability to mimic currency. If alchemists were not really able to transmute species, they were able to fake it, but none of the ritual dangers attendant to magic in *Super illius specula* reflect on alchemy.[95]

The description from *Pro conservanda* fits perfectly into this context, in spite of the rather unsophisticated method of forgery it addresses. The most likely explanation for opening the section on gold and silver with a discussion on forgery is that what follows has very little to do with medicine and health. Instead, the text then picks up from Bartholomaeus's *Properties* and moves on to a discussion of gold's ability to retain its weight in a furnace while being purged of its impurities. Likewise, gold placed in a furnace will not evaporate, nor be reduced in weight, but however much impurity or dross is mixed in, it leaves that in the flames, and becomes more and more pure.[96] Intriguingly, a passage from *Properties* on the making of alchemical gold from sulfur and mercury is omitted, though a parallel passage on the role of quicksilver in the making of other metals is included later in an entry on "silver and metal (*argento et metallo*)."[97]

Speculation aside, we know from *Pro conservanda* (and from *De proprietatibus rerum*) that gold is the preferred treatment for leprosy.[98] *Pro conservanda*, however, also dwells on gold's palliative properties for melancholics and those suffering from defects (*vitia*) of the spleen.[99] What is left is a rather sparse collection of the healing properties of gold and a few alchemical tidbits. Given that the medical and alchemical community held gold in great esteem, the slightness of this entry might be explained by relatively little access to sources beyond *Properties* or (more interestingly though no more likely) due to some concern about demonstrating alchemical theory. The opening sentence makes it clear that the author at least was aware of the concern over forgery, even if the passage itself was intended only as a transition to gold's essential characteristics.

The dependence of *Pro conservanda* on *Properties* is also apparent in the section on planets. Where Bartholomaeus offers an entire book on celestial bodies and constellations, *Pro conservanda* limits itself to one short entry. Though it omits the effect of planets on metals contained in the *Properties*, it does paraphrase the encyclopedia's concise explanation of astrological effects, which conforms to the basic tenets of astrology discussed in the prior section.[100] *Pro conservanda* goes on to list the properties of a few prominent celestial bodies, but skims over—nearly to the point of eliding—direct discussion of the celestial effects on human beings, though such a connection would have been assumed by the reader. Given that the discussion nearly skips over astrological determinism (not to mention astrological medicine) as well as the movement and composition of the heavens, one could surmise that the author chose to avoid discussion of these matters. At the same time, there is no positive evidence for such a decision, especially if one considers the elision here as part of a pattern of summarizing and sampling *Properties.*

Both *Pro conservanda*'s and *Properties*' sections on "Waters" are quite long, though in this case the section in *Pro conservanda* is lengthier.[101] The author of *Pro conservanda* has lifted entire passages from throughout the prior encyclopedia, however.[102] The discussion ranges widely, delving into the nature of rain, lakes, seas, and fountains. The bulk of the entry, however, is really more about what we would term liquids, and their various palliative and deleterious effects, such as sulfurous water, barley wine, and the right types of water for boiling beans.[103]

The key passage that concerns us, however, has no parallel in *Properties.* Here, the author of *Pro conservanda* turns to a discussion of "artificial waters (*aquarum artificialum*)," which includes the aforementioned barley water as well as wine.[104] Greatest among these human-created waters, however, is the *aqua ardens* (literally, burning water), which derives its name from the fact that "if it is spilled over the hairs of the head or cloth and touched to a flaming candle, it immediately catches fire, and the hair or cloth will seem to burn, however the cloth or hair is consumed not at all, but the flame endures for a long time until that water is consumed by the flame."[105] It is important to note that the quality of the *aqua ardens* was that it burned without consuming something other than itself.

This property can be observed in highly distilled alcohol, and this is precisely what the author of *Pro conservanda* instructs the reader to make. "Take good wine, and strong, pure, and red and place it in an alembic, and distill it over a light fire. The water will become rosy, and *aqua ardens* will emerge through sublimations (*sublimationes*). And if it is distilled often, however much more it is distilled, so much more refined and useful will it be."[106] The language of alembic and sublimation is telling here, for it speaks to a certain expertise not available to encyclopedists, but rather to those who had had some encounter with alchemy. The precise, if simple, discussion of distillation may not be useful enough to actually instruct someone in the process, but it demonstrates the author's knowledge.[107] A later passage adds to what we know of the author's expertise, when a number of further alchemical uses to which one can put the *aqua ardens* are listed: "It freezes mercury, whitens copper, dissolves calcified bodies and spirits."[108] This is a reference to the process of and components used in aurification without any specific mention of gold itself, revealing the author's understanding of metallurgical alchemy while maintaining some distance from suspect practices.

The history of *aqua ardens* is quite old. C. Anne Wilson theorizes that the Latin *aqua ardens*, or fiery water, was the same as the ancient *theion hudōr*, or divine water, from ancient Greek alchemical texts, whose primary property seems to have been its flammability. Wilson describes the earliest concoctions of divine water as a kind of sulfur water or mixture of water and quicklime having religious uses in the ancient world.[109] The earliest record of *theion hudōr* as a distillate, however, occurs later, in the text of Pseudo-Hippolytus's *Refutationes omnium haeresium* (*Refutations of every heresy*), even though afterwards the recipe lay dormant in the West for hundreds of years.[110]

Wilson notes that not long after Michael Psellos, the eleventh-century Byzantine historian, documented the continued existence of *theion hudōr*, there begins to be evidence of a similar product in the Latin West after an absence of such mention since late antiquity. For instance, among the earlier alchemical texts that include an *aqua ardens* recipe is the *Ars alchemie* attributed to Michael Scot. The *Ars alchemie* primarily concerns itself with transmutation of materials, namely metals, but it does mention *aqua ardens* as a possible substitute for turpentine, reiterating its flammability.[111] In this case turpentine or *aqua ardens* is combined with sulfur as the ingredients for a primitive flamethrower, which naturally is shaped like a metal man or animal so the flames can gout from a mouth. This product is similar to the antique *theion hudōr*, and is typical of other early *aqua ardens* texts, where fiery water is not medicinal.[112]

The nonmedicinal tradition of *aqua ardens* is present here, in what is almost certainly a tacit reference to making alchemical gold. In explaining how to use the *aqua ardens* to make an acid (probably nitric acid, given the use of saltpeter) that "dissolves or liquefies every metal," including steel, silver, copper, lead, gold, and others, the author also notes that the same acid if dropped lightly on a cloth gives it the color of gold.[113] As with many entries, the passage tends toward the encyclopedic rather than the systematic. A listing of qualities and uses of the fiery water is somewhat lacking in additional detail, and it would have been notable had the author continued to discuss the ramifications of *aqua ardens* for metallic alchemy. Still, the treatment here of the *aqua ardens* does foreshadow the qualities ascribed to it by John of Rupescissa a half century later, where its healing powers come to the fore.

According to the author of *Pro conservanda*, *aqua ardens* "is reported to have forty virtues or effects."[114] The range of applications is staggering, and the curative properties of the distillate outnumber its uses for metallic alchemy. Some effects are banal, though undoubtedly invaluable in the Middle Ages: "It takes away stench from the nose and gums and armpits."[115] Other effects are those one might expect of distilled alcohol: "It destroys skin ulcers if smeared with it many times, and internal ulcers, if drunk. It takes away spots of the eyes, and their redness and heat . . . makes a person cheerful about anything . . . cures salty phlegm, blemishes and rosacea, removes pain from teeth . . . [and] destroys abscesses of the throat if frequently gargled."[116] It is, of course, recommended to be imbibed by those who are faint of heart (*pusillanimis*) or pregnant (*praegnans*).[117] There was a limited tradition of *aqua ardens* being used as a curative prior to the thirteenth century, anticipating many of the uses in *Pro conservanda*. There are, for instance, early eyewater recipes, the most famous of which are those attributed to Peter Hispanus (1215–77), later Pope John XXI.[118] Eyewaters, used to treat scores of eye problems prevalent in the Middle Ages, had ingredients and methods of production that were quite similar to those used to make the *aqua ardens*. There is also a thirteenth-century recipe for *aqua ardens* from Weissenau that claims the substance can be used for contraception and abortion.[119] Still other effects of the *aqua ardens*, however, are quite surprising; it "cures wounds . . . [and] cures paralysis, when it is frequently applied to paralyzed limbs. It sharpens the intellect if taken temperately, recalls the forgotten to memory . . . preserves youth and retards age . . . greatly comforts the melancholic . . . and cures dropsy at the source."[120] It cures cancer, halts leprosy, and destroys kidney and bladder stones. It also dissolves cataracts and heals the deaf.[121] Although in most cases the remedies supplied in *Pro conservanda* are naturalistic, these remedies hint that the *aqua ardens* is something more powerful than its production would suggest. As Bacon discussed, one of the chief powers of the elixir was to retard aging, an effect here of *aqua ardens*. In this case, however, *aqua ardens* resembles the theriac more than the astrologically derived elixir of Bacon.

The theriac was a naturalistic compound, sometimes including nearly 80 ingredients.[122] It was championed (though not invented) by Galen, and it had an enthusiastic following at Montpellier by the late thirteenth century.[123] The main ingredients of the classical theriac—or at least the most striking—were viper flesh and sometimes opium.[124] While Galen and others conceived of the theriac as a universal remedy, one of its chief functions was to ward off the effects of poison, either as an antidote or as an inoculation when consumed systematically.[125] Thus, one of the first clues that *aqua ardens* of *Pro conservanda* is related to theriac is that it is an effective anti-venom and that "its odor kills reptiles (*odor ipsius reptilia interficit*)."[126]

The rationale in the late thirteenth century for the effectiveness of theriac follows Avicenna's understanding that the whole of the theriac was greater than the sum of its individual ingredients. McVaugh notes that the potency of the theriac was "*sui generis*, arising out of the act of combination."[127] In the Middle Ages, physicians such as Arnold of Villanova had taken to classifying "theriacal medicines," regardless of whether snake parts were present.[128] The more general theriacs employed were medicines compounded of various spices and herbs, made into a syrup (hence the English derivative, *treacle*) in which serpent flesh was not a featured ingredient.[129] Here again we see a parallel in the *aqua ardens* of *Pro conservanda*, where nutmeg, clove, ginger, cardamom, and cinnamon, among other spices, are to be added to the *aqua ardens* in order to make an effective curative.[130] Thus, many of the healing properties come not only from the properties specific to the distillate ("it extracts the power of all herbs if they are placed in it"), but also from the conjunction of the distillate with various ingredients.[131] Finally, the *aqua ardens* of *Pro conservanda* shares some of the theriac's "constrictive" properties.[132]

There is no evidence to suggest any kind of specifically religious discourse influencing the elixir described in *Pro Conservanda*. The effectiveness of the *aqua ardens* in working metals and curing people certainly speaks to the elixir tradition, but, apart from adopting the *aqua ardens* into the tradition, there is nothing particularly innovative. No doubt this is due to the broad reliance on the work of others. It can be difficult to innovate when you are copying. The fact that *aqua ardens* is added to material copied from Bartholomaeus, however, is instructive, as it stands opposed to the many excisions in regard to metallurgical alchemy. If anything, it makes those excisions all the more conspicuous, given the alchemical expertise required to work with *aqua ardens.* If the text was compiled by Vitalis of Furno, it was probably done before he earned his red hat in 1312. Vitalis was a cautious and conservative politician and thinker in a troubled time for Franciscans. Moreover he was operating in an era in which alchemy was coming under greater suspicion from Franciscans and other ecclesiastical authorities. We cannot know what kind of text Vitalis might have written had these constraints not been present, but *Pro conservanda* demonstrates the way religion might constrain alchemy—not necessarily in a grand sense of denying its reality, but in a way analogous to other constraining political or social discourses, which might marginalize practitioners of a marginal practice. The example of Vitalis is an important reminder that religion is not limited to doctrinal or theological thought. Yet, as the case of Vitalis suggests, ecclesiastical distrust or censure of a practice does not lead to its end. There remained a way of talking about alchemy without directly addressing or championing its practice. A more striking example of the failure of ecclesiastical suspicion to constrain alchemical discourse—though it did shape it—emerges in the case of John of Rupescissa.

John of Rupescissa and the Quintessence

Rupescissa's contemporaries considered him a practitioner of the *scientia* of alchemy, a conceit shared by modern scholars.[133] Rupescissa, however, objects to this categorization. The terms alchemy and alchemist, in fact, hardly surface at all in his major "alchemical" work, *On the Consideration of the Quintessence* (*De consideratione quintae essentiae*), and when they do they are objects of scorn.[134] The avoidance of the label of alchemist mirrors John's refusal of the mantle of prophet (*propheta*), a title also bestowed upon him by some of his contemporaries.[135] Whatever he says, both prophecy and alchemy are practices in which he engages himself with vigor. I will return to John's designation as prophet in the next chapter, but his rejection of the label of alchemist is, at least in his own terms, fully warranted. John also objects to the term "natural philosophy," which he pointedly labels "worldly philosophy (*mundana philosophia*)."[136] The creation and use of the elixir is, in John's own eyes, an act of devotion. However much he relies on natural philosophy and discusses alchemical "recipes," when dealing with the object of his alchemical quest, the creation of what he calls the "quintessence," a distinction arises. What has heretofore been a recipe becomes simultaneously an activity that comes closer to a ritual. John's discussion of the elixir as "our heaven" or "human heaven" is no mere code word.[137] This terminology suggests that the making and use of the elixir privileges meaning and connection to the divine. The idea that there is a heaven inside each person that resonates with the greater heaven of the cosmos is, for John, an illustration of the world as it could or should be. In tapping into the subjunctive world of ritual, John is perhaps correct to reject the title of alchemist. For though John's work belongs to the alchemical tradition, the means by which he transformed alchemical practice into a vehicle for approaching the divine also sets him outside that tradition.[138]

In 1327 John of Rupescissa enrolled at the University of Toulouse. His first five years were a tumultuous time, and he tells his readers that he was caught up in the quest for fame and the vanities of the world. It was during this period that he read widely, delving into whatever subject struck his fancy.[139] Not surprisingly, it was also during his first years at university that he encountered the subject of natural philosophy, that is, alchemy, though it would be many years before these early studies would bear fruit in his two alchemical works, the *Liber lucis* (1350), and the much larger *De consideratione* (1351/2).[140]

John's alchemical works were written while he was under house arrest at the papal court at Avignon. He had been accused of heresy, but such claims were ultimately found wanting. The curia nevertheless extended John's captivity on the grounds that he was a *phantasticus*, that is, mad, lying, or being deceived by demons (or possibly all the above).[141] Without rehearsing all the events of John's life, it is important to reiterate for the purposes of this discussion that John's alchemical works were written after he was declared a *phantasticus*, as well as after he had earned some measure of notoriety for his prophetic works. The shift to alchemical writing did not signal an end to John's strictly apocalyptic work, which continued throughout the 1350s. Apocalyptic anticipation played a significant role in the development of John's alchemical texts, and will be discussed in Chapter 3.

The focus here is John's alchemical magnum opus, *De consideratione*.[142] Throughout this section, preference is given to the term *quintessence*, which connotes the idea of perfection. The notion of the quintessence being, in fact, a fifth essence—a substance outside the four elements, fire, air, water, and earth—is extremely important to the ideation of John's elixir.[143] In John's work, the quintessence is synonymous, however, with *aqua ardens*, the same substance found in *Pro conservanda*. There are some significant differences between the two, as well as from prior notions of both quintessence and *aqua ardens*. While John posits a specifically Christian cosmological frame for his alchemy and considers it divine knowledge, he still adheres to Aristotelian natural-philosophical principles, even if the Philosopher goes unnamed.

De consideratione is a collection of alchemical processes punctuated by theological musings, apocalyptic predictions, and self reference; the latter sometimes veiled with humor and wordplay, sometimes transparently upset and frustrated.[144] Formally speaking, *De consideratione* is a work divided into two books. The first and longer book describes the characteristics of the quintessence through an examination of its heavenly origins and, ultimately, identification of the quintessence as synonymous with *aqua ardens*.[145] John also explains how the quintessence affects humanity by illustrating various properties of astrological medicine.[146] Next, John elaborates on the methods of extracting the quintessence.[147] The last section of the first book is a very long discussion of how the quintessence merges with and amplifies various properties of other medicinal substances. Book Two is much slimmer, but has the advantage of illustrating how the quintessence or *aqua ardens* can be used to treat specific ailments, whether their etiology is natural, astrological, humoral, demonic, or some combination thereof.

In *De consideratione*, Rupescissa theatrically reveals that the miraculous cure-all, the quintessence, known to him through divine revelation, is in fact an already known alchemical compound: the *aqua ardens*, also known as *aqua vitae* (the water of life).[148] This makes a certain amount of logical sense in Rupescissa's understanding of the quintessence, as we shall see, but fiery water and quintessence had long had existences independent of one another. These histories are well chronicled, so for the purposes of understanding the quintessence-cum-*aqua ardens* as John conceived of it, a brief outline of each will suffice.[149]

Aristotle introduced the notion of the quintessence as a heavenly element separate from the traditional four elements (earth, air, fire, and water), though it was later thinkers who would actually name it as the fifth element. Unlike Plato in his cosmology, Aristotle conceived of the heavens in his *De caelo* as "eternal, perfect, and divine."[150] This view quickly overwhelmed the Platonic "model" found in *Timaeus* that asserted celestial bodies to be mainly composed of fire.[151] By the thirteenth century, celestial perfection was the dominant scholastic view, as was its corollary that the heavens were composed of a fifth element (*quinta essentia*).[152] For Aristotle, however, this substance beyond the traditional four elements was confined to heaven, and did not mix with earthly elements.[153] Aristotle further opined that while it was not the quintessence per se that was responsible for the light of the sun and stars; it was the proximity of concentrations of the quintessence to the celestial bodies that gave them their luminescence. The important distinction made by Aristotle was that heaven or heavenly bodies were not, in fact, fiery objects, arguing against "the presumption that a thing is composed of the same stuff as that in which it is situated."[154]

Aristotle's most influential mediator in the Middle Ages, Thomas Aquinas, did not entirely embrace the cosmology of the Philosopher. When dealing with Aristotle's contention that there exists a perfect substance beyond the four elements, Thomas notes that he argues "unfittingly (*inconvenienter*)", and even resorts to punning as a means of turning aside Aristotle's proof.[155] Thomas's aim, however, is not to disprove Aristotle's thesis, but to correct

Aristotle's claim that heaven and heavenly bodies are outside nature.[156] That said, Thomas does not dismiss Aristotle's notion of something outside the four elements, for he does concur that elemental forces are finite and changeable, two qualities wholly inappropriate for heaven. This point is made clear when Thomas takes up Aristotle's argument on the nature of the stars, namely that they are perfect because they are part of the heavenly sphere.[157] Thomas raises a number of objections based on what one might describe as common-sense observation of the heavens, then deals with each of them logically in order to bolster the argument of the Philosopher. The primary issue he must deal with is that heavenly bodies appear diverse, which would imply that they are contrary or, at the very least, mixed.[158] The second issue is that many celestial bodies, the stars in particular, do indeed seem to be made of fire. If either of these objections had merit, however, then heaven would be corruptible, which was clearly impossible.[159] Thomas argues that, in the first case, "properly speaking, not every difference is a basis for opposition (*contrarietas*)."[160] The stars might seem different, he says, but are not. The difference in luminosity is a question of the relative density of the heavenly matter contained within each celestial body.[161]

In regard to the alleged fiery nature of stars, Thomas restates Aristotle's premise that heat and light are due to friction caused by the motion of the upper bodies. This, he concludes, accounts for heat, but not light, so he provides two solutions to the problem. He cites Alexander of Aphrodisias, who writes that Aristotle solved this problem in *De anima*, where the Philosopher states that light is common to both fire and the heavenly substance.[162] Yet Thomas also contends that Aristotle's argument in *De caelo* suffices, for it provides that friction causes fire in the lower bodies, accounting for both heat and light.[163]

To be clear, there is nothing in Thomas's commentary that suggests he would have supported John's claim that the fifth element could be distilled as an alchemical compound, or even found on earth. Nor does John cite Aristotle, Thomas, or any other commentator in his own argument for heavenly perfection. Yet Thomas argues for three cosmological notions essential to John's alchemical undertaking. The first is the Philosopher's argument that heaven is perfect and immutable or impassible. The second is that the heavenly fifth element is *natural*, and not outside or contrary to nature, as Aristotle seems to say.[164] The third notion is that the fifth element is composed of matter, which may have been the most controversial point on which

John's argument depends.[165] Thomas differed with Averroës on this point, and many scholastics continued to attack the notion throughout the fourteenth century.[166]

Though Thomas insisted that the heavenly substance was composed of matter, he concluded that it was different from terrestrial matter. Likely closer to John's own understanding of the quintessence were the ideas of Giles of Rome, and John's fellow Franciscan, William of Ockham. Ockham's discussion of heaven probably comes closest to John's, in that he conceives of heavenly matter as incorruptible, save for God's will. As Edward Grant points out, by introducing the possibility of celestial corruption by God, even if it would never happen, Ockham leaves open the idea that the heavenly substance is subject to manipulation, a theory which concords with John's alchemy.[167]

While many of the most famous scholastics who were busily integrating Aristotle into the study of theology also mediated new knowledge on natural philosophy, scientific learning in the Middle Ages relied on spurious as well as genuine contributions of antique and Islamic authors. In an early thirteenth-century text once ascribed to Robert Grosseteste, *De generatione stellarum*, the Aristotelian notion of the fifth essence is combined with the alchemical claim that the quintessence can be produced: "Alchemical doctors suppose that inside every natural mixed body there is a *quinta essentia*, like something that encompasses all four elements."[168] This idea accords with Rupescissa's text as well.[169] The authentic Grosseteste in his *De cometis* also discusses the quintessence as *res corporeae spirituales* ("corporeal-spiritual things"). For John, the quintessence was firmly a corporeal thing and a part of the natural world. Its heavenly origin as a perfect and unchangeable substance made it no less a substance. Michela Pereira argues that this medieval relocation of the fifth element "effaces" the difference between heaven and earth.[170] For John, it might be better to speak of bridging heaven and earth, though the bridge, so to speak, goes in only one direction. Since it is a heavenly substance, John's text asserts that obtaining the quintessence takes not only alchemical skill, but also the grace of God, though he often contradicts himself on this point. Obtaining heaven on earth, to borrow Michela Pereira's phrase, is not only a distillation, it is also a miracle—an idea I will take up in more detail at the end of this book.

The development of *aqua ardens* as a medicinal substance was still a relatively recent innovation when John adapted it for his own uses. While Rupescissa was the first to medicinalize the quintessence, the *aqua ardens* in its form as a distillate had a prior history of medicinal use, as described above. Nevertheless, John's ascription of the quintessence as the chief substance of *aqua ardens* does seem to be innovative, as do the reasons for its special curative properties.[171] John treats the *aqua ardens* exclusively as a distillate, synonymous not only with the quintessence, but also with *aqua vitae* (water of life) and *anima* or *spiritus vini* (the soul or spirit of wine).[172] Spirits, by which name we still call distillates, was precisely what John was making and calling the quintessence. Much of the text describes the method for creating a still as well as for careful distillation of alcohol—using only the best wine, of course—by a process of evaporation and condensation.[173] In higher levels of purity, the distillate, which John refers to as the quintessence or *aqua ardens* with fairly equal frequency, proves to be something that cannot be ascribed to the interaction of the four basic elements of the universe: fire, air, water, and earth.[174] It cannot be hot and humid like air, because it prevents spoilage (such as the generation of spiders and flies on material contained within it). It cannot be water, since it can ignite. It cannot be earth, which is cold and damp, because it warms you when you drink it. And, finally, it cannot be fire, because it cools fevers and objects placed within it.[175]

The secret of its nature can be found in its preservative properties. "God created it for the preservation of the four qualities of the human body, just as he created heaven for the preservation of the entire universe."[176] John holds that within the human body, and every substance, is a part of the heavenly substance. This is granted to humans and to other created things by God through the sun and the stars. Here John turns to astrological medicine to prove that the "adornments of heaven"—the sun, stars, planets, and moon—are not fiery objects. Otherwise they would belong to the four elements, and thereby be changeable and of no use to the creation of a preservative medicine.[177] Instead, celestial bodies had to be impervious to flame and any sort of corruption or consumption. What is more, they provide us the quintessence with which we can adorn and preserve our own bodies. The mechanism by which this occurs is related to the astrological theories discussed above; that every heavenly body or constellation has a particular purview. In astrological medicine, one can break down the influence of the stars by body part or even type of ailment.[178] Most important for John was that this same theory had been proposed in an earlier treatise on *aqua vitae* by the noted Catalan physician and Joachite, Arnold of Villanova, with whose authentic and spurious works John was acquainted.[179] According to John, the quintessence provides the link between bodies celestial and terrestrial and allows for the divinely mandated effect of sun and stars over people and things. Therefore, reasons John, by augmenting the amount of quintessence in a given healing herb or stone, one can similarly augment its effect. Thus, an herb like hierapicra, which has the ability to heal the head and upper parts of the body, becomes far stronger when used in conjunction with the quintessence.[180]

The quintessence, then, can be used to heal any imagined illness or wound, thanks not only to its power of amplification, but also because when distilled in various ways with various ingredients, it has the ability to balance the four qualities or humors of the human body; it "would cut off every excess of some quality, and replenish whatever quality was lost."[181] In the humoral model prevalent in medieval medicine, this was a potent effect indeed. The quintessence does not, however, grant immortality. Like Bacon before him, John implies that his elixir could grant immortality, save that such an effect would be contrary to divine command. Instead, the elixir preserves the body until the day of death preordained by God. John reasons that God has something like an expiration date for every human being. He has not ordained the day of death exactly, else why worry about a life-saving medicine, but rather something like a last possible day of life beyond which one cannot pass.[182] The elixir, however, is proof against death via illness or corruption of the body and likely against accidents that impair the body, but do not kill it outright. John, however, fails to prescribe how one determines such a premature death from the final end, leaving the limits of the quintessence less distinct than his avowal suggests.

* * *

What I have tried to demonstrate in this chapter is the variety of ways religious convictions—those of the authors as well as those of the papacy or curia—have shaped the pursuit of the elixir. While identifiable alchemical genealogies are present, those genealogies on their own cannot account for the specific ways the authors above theorized their elixirs. Rather, we can see in these authors the ways Franciscan approaches toward the created world manifested themselves in the development of the elixir tradition. Sustained meditation on the natural world was part of Franciscan spirituality, which, in turn, meant that Franciscan spirituality could well be imparted to natural philosophy.

Religious discourses were both constructive and coercive—sometimes at the same time. In the cases of the inestimable glory and the quintessence, the potency of each substance was such that they required a connection to the heavenly realm in order to account for their amazing powers over the human body and psyche. These were no mere medicines. At the same time, the exigencies, policies, and suspicion of other clerks also shaped the presentation of the elixir. This is not limited to the possible work of Vitalis of Furno, though *Pro conservanda* has served as the principal example of this idea in the current chapter. As we shall see in the next chapter, however, the exigencies of the rapidly approaching end times and the impending battle with Antichrist also served to shape and justify the elixir. While not all three authors agreed that the needs of sacred history impelled the creation of the elixir, they all understood the approach of the end as justifying a particularly Christian reading of the natural world.

CHAPTER 3

The Apocalyptic Imperative

Thus far, we have seen how alchemy existed against the backdrop of Franciscan philosophical speculation regarding the created world and the various religious and philosophical traditions that made up the genealogies of the elixirs. I have omitted until now what is at least a curious coincidence, but more likely a meaningful intersection of discourses. Each proponent of the elixir—Roger Bacon, Vitalis of Furno, and John of Rupescissa—authored at least one text dealing with the end times. I have separated the discourse about the apocalypse because, while it certainly inflects the narrative tone of the alchemical texts, it has a very different genealogy. Speculation about the end times was widespread among the Franciscans, and many brothers wrote formal commentaries on the Book of Revelation.[1] Yet wrestling with the apocalypse was a weighty, sometimes visceral concern for each of these authors. In this chapter, I discuss and describe their apocalyptic commitments and whether these are something more than incidental to their understanding of the elixir. Can we disentangle the apocalyptic discourse from the alchemical, and, if so, what are the consequences?

Franciscan apocalyptic has remained a popular subject among medieval historians for more than half a century, due no doubt to the conspicuous and radical apocalypticism that emerged within and around the Franciscan Spirituals. It was Leah DeVun, however, who in her monograph on John of Rupescissa first made a convincing argument that Franciscan alchemy and Franciscan apocalyptic ought to be read together. Rupescissa's use of Christian "metaphor," she argued, was "central to the way in which Rupescissa formed and communicated his alchemical theories."[2] One of the key contributions of DeVun's study emerges from her attention to language, especially the way in which Rupescissa employed alchemical language to discuss the end times. While the kind of linguistic play in which Rupescissa engages is not shared by Bacon or Vitalis, this chapter builds on the idea that each of these authors' alchemical works ought to be read with some attention to the larger contours of their oeuvre, especially when the elixir or alchemy plays a role in the drama of the end times. The work of this chapter, however, is to push these arguments farther to illustrate a very basic connection between the elixir and apocalyptic expectation that has not been explored: namely, that both seek to realize spiritual truth in a material way.

Caroline Walker Bynum has written that in the period after 1100, Christianity became paradoxically interested in materiality. That is to say that interest in the material grew, somewhat irreconcilably, at the same time as Christians increasingly focused on interiority. Bynum's argument focuses very much on the material aspects of lived Christianity in the later Middle Ages, such as devotional art, sacramentaries, and eucharistic miracles, as well as doctrines such as the Incarnation that attend to materiality.[3] I would argue that apocalyptic thought is not far removed from material concerns. Bynum herself notes the importance of the Resurrection, a key event of the apocalyptic drama, and I would add that apocalypticism overall suggests a kind of yearning for the worldly realization of abstracts—good, evil, salvation, damnation. And, as suggested in Chapter 2, the production of the elixir also includes the idea of making heaven tangible. While these two efforts to realize the heavenly are quite distinct, they can be mutually reinforcing.

The apocalypse was not necessarily an event to be dreaded. It offered for the faithful, or at least the zealous faithful, the opportunity to demonstrate their faith in ways not possible since the early Church. Many would have welcomed martyrdom at the hands of Antichrist. What happened after Antichrist's fall was a matter of debate—but living through the events of the Book of Revelation made one party to an epoch that rivaled the first coming of Christ in import. What is more, Franciscan apocalyptic thought grew from the roots of Joachim of Fiore, who, contrary to traditional views of the end times, argued that two groups of people—two orders of new spiritual men, he called them—would be the vanguard against Antichrist. Many Franciscans considered their order one of those prophesied by Joachim. Engaging in conflict with Antichrist and his servants changed a struggle for salvation that was largely spiritual into one that was rooted in the material world against a living foe.

Toward Franciscan Apocalyptic

Of the many narratives that emerge from the first century of the Franciscan Order, one of the least expected may have been how quickly a significant number of the Friars Minor became attached to apocalypticism. This adoption of apocalyptic thought occurred in the context of rapid, one might say unchecked, growth. The Franciscan Order was less circumspect regarding its membership than were their rivals, the Dominicans. Within Francis's own lifetime, what had been a small band of devout lay penitents had grown into an international conglomerate that contained in its ranks ordained priests, schoolmen, and lay persons, whose coexistence was not an easy one as they contested over the legacy of their still living founder. Francis himself was uneasy about the shape his Order was taking, having ceded governance of his brothers well before his death to men he believed more capable of coping with the bureaucracy necessary to deal with far-flung brethren and incipient papal demands.

Francis drafted a Rule in 1221, another Rule in 1223, and a *Testament* shortly before his death in 1226, in large part because of the rapid development of the Order and its growing intellectual clout. Francis himself was, if not hostile to secular education, certainly wary of it, and did not think that preaching repentance required educated men.[4] In his *Testament* Francis expressly forbade glosses on his Rule and even condemned any brother who might add to or subtract from his *Testament* itself.[5] Of these prescriptions and proscriptions, however, only the 1223 Rule gained papal sanction, and in 1230 Pope Gregory IX issued the bull *Quo elongati*, which rendered the strictures of Francis's *Testament* non-binding.[6]

The first life of Francis, written by Thomas of Celano in 1228, reflects the continued angst regarding the growth of the Order. In response to the uses to which the Order is being put, Thomas reports that Francis feared that the Order would devolve in the future, to the point that brothers would be like "inedible fruit," though they will appear to be devout.[7] Thomas's second *vita* of Francis, composed in 1244 at the behest of Crescentius of Iesi, then Minister General of the Order, demonstrates growing conflict between rigorists and those who had accommodated themselves to the demands of the *Ecclesia*.[8] At the same time, many within the rigorist wing of the Order had grown enamored with the exegetical strategy and apocalyptic predictions of the twelfth-century abbot, Joachim of Fiore.[9]

After Joachim's death, commentators regarded him as a prophet of Antichrist, but his impact on the medieval imagination far exceeds that of titular prophet.[10] The Franciscans especially, but also to a lesser extent the Dominicans, felt the influence of Joachim. And even those who did not agree with Joachism nevertheless felt compelled to respond to it, so persuasive were Joachim's methods and predictions. Joachim's own order, the Florensians, were not a transnational group like the mendicants, but they too continued to ensure the Calabrian abbot's influence was felt in southern Italy. Lay groups, including various beguines in southern France, as well as the apostolic brethren under Gerardo Segarelli and then Fra Dolcino, were outlets for more radical interpretations of Joachim's predictions.[11] And centuries later, Protestant reformers such as John Bale appropriated Joachim for their own apologetic ends in the wake of the Reformation.[12]

The difference between Joachim's understanding and the traditional apocalyptic view manifests itself in a number of ways. Foremost among the novelties he introduced to apocalyptic exegesis and expectation was his scheme of Trinitarian history and *concordia.*[13] Since a basic understanding of this view is necessary to understand the more concrete predictions that emerged from Joachite speculation as well as the ways in which Roger Bacon, Vitalis of Furno, and John of Rupescissa reacted to it in their writings, I will restate some of the discussion in Chapter 1 before elaborating on the apocalyptic dimensions of Joachism.

Trinitarian history emerged from Joachim's exegesis of scripture. Joachim accepted the traditional typological model that understood the events and figures of the Hebrew Bible to prefigure those of the New Testament. Yet he added to it a third typology in which both testaments prefigured events and persons after the period of the New Testament. This relationship of the books of scripture he called *concordia.* Joachim understood salvation history to unfold in patterns, especially patterns of two, three, seven, and twelve, all of which, when layered over one another, expressed the fullness of salvation history. The pattern of twos related to the two dispensations, the two covenants, the two Testaments, the Alpha and the Omega. Layered over this pattern of twos is a pattern of threes, represented chiefly by the Trinity. Each member of the Trinity was the guiding force over one *status* of history. The Latin *status* is often translated as "age," but this is insufficient to express the fullness of Joachim's meaning (a *status* was not merely a span of time, but had specific attributes or characteristics that defined one from another).[14] Joachim also separately used the term *etas* (age) in his pattern of sevens. The first *status* belongs to the Father, the second to the Son, and the third to the Holy Spirit.[15] Joachim believed, to put it as simply as possible, that the world to date had existed in two separate ages that corresponded with the two Testaments. First was an age of patriarchs, the time of the Old Testament, when God the Father ruled humankind through the commandments and the laws of Israel. The second age was that of the Son and the New Testament, when the precepts of Jesus Christ governed the Church. Thus, in contemplating the mystery of the Trinity, Joachim prophesied that surely a third age, that of the Holy Spirit, must follow the *status* of the Son and was, in addition, imminent. During this *status*, understanding of the revelation given in the Gospel message would achieve new heights of clarity and profundity unseen even in apostolic times.[16]

Believing that the pattern of threes remained consistent with the pattern of twos, Joachim did not think that the third *status* would usher in a new dispensation, a new covenant, or a new Church. The *synagoga* (synagogue) and the *ecclesia* (church) accounted for the pattern of twos, meaning that the *ecclesia* would survive into the third *status*. In the third *status*, however, the nature of the *ecclesia* transforms from the *activa* (active) to *contemplativa* (contemplative).[17] The third *status* changes the essential nature of the Christian experience, which becomes characterized by a richer and fuller understanding of scripture—a shift of emphasis, but only emphasis, from the apostle Peter to the apostle John. Joachim was clear, in a way many of his inheritors were not, that the *status* of the Holy Spirit would not abrogate in any way the institutions of the second *status*—the Church, Petrine (papal) authority, and the New Testament.[18]

Joachim's pattern of history called for a new understanding of Antichrist, whom Joachim, echoing some patristic sources, interpreted as a twofold or multiple enemy.[19] Joachim drew the final enemy in one of his most famous *figurae*, the drawings he made to enhance and explain the exegetical truths he inferred through *concordia*. In the *figura* of the dragon of Revelation 12:13, Antichrist is represented twice, as both the seventh head and the tail. As the seventh head, he represents the "seventh king who is rightly called Antichrist," the last and worst of a series of enemies of the *ecclesia* during the *status* of the Son.[20] This Antichrist is prefigured by such personages as Herod, Nero, Mohammed, and Saladin.[21] It is important here to point out that Joachim's chronological patterns meant that Antichrist was a recurring presence in the world. Evil personified itself regularly and, as Joachim's list suggests, warred against Christendom in a very real sense. The recurring embodiment of evil culminates in the tail Antichrist, the final enemy of the last *etas* and the third *status*, called "Gog, the final Antichrist."[22] Joachim was concerned by the final Antichrist, but the Antichrist of the tail was not an immediate threat, and would represent the power of Satan loosed on the world only after the thousand-year Sabbath. Rather, Joachim referred to the seventh tyrant as the "greatest" or "great" Antichrist, despite being a mere human. One other key difference in Joachim's conception was that the great Antichrist would not be a Jew (as the traditional Antichrist narrative stated), but a false Christian (whose allies nevertheless were similar to those depicted in normative Antichrist legends).[23]

Another change Joachim made to the standard legend of the end times was the omission of Enoch and Elijah. In the traditional narrative, the reappearance of the Jewish prophets Enoch and Elijah occurs as a counter to Antichrist's "purge" of Christendom. Enoch and Elijah were viewed as typologically foreshadowing Christ, and according to scripture, neither died.[24] Antichrist would kill both, but leave them unburied, allowing them to rise and make their heavenly ascent after three days. Their brief ministry will rally many to true faith, and their appearance will foreshadow the end of Antichrist's reign. Joachim adapted this idea, but understood Enoch and Elijah to point to two separate groups of *viri spirituales* (spiritual men), who would assume the duties of the two prophets during the end times.[25] Joachim described one group of spiritual men who were dedicated to preaching and another to contemplation and prayer. Many observers considered this prophecy to have foretold the rise of the mendicant orders. Many Franciscans eagerly agreed, offering themselves the possibility of being critical actors in the battle against Antichrist. Some Dominicans did as well, though less vociferously (and less disastrously) than the Franciscans.[26] These groups of spiritual men, under the direction of the pope, would actively, and successfully, challenge Antichrist and his followers.[27]

Joachim's identification of *new spiritual men* as papal allies and foes of Antichrist eschewed elements of the Antichrist legend in favor of a characterization of the Christian *ecclesia* as active, even defiant, in the face of Antichrist's oppression. Bernard McGinn astutely notes that "Joachim's writings were the first to give the papacy a special role, both in the coming crisis of the present era (the *secundus status*), and also in the future terrestrial triumph of the Church that the abbot spoke of as the *tertius status*."[28] Joachim's idea of the papal role is not limited to fighting Antichrist: "There shall ascend . . . a universal pope of the New Jerusalem, that is of the holy mother church. . . . full liberty shall be given to him to renew (*innouandum*) the Christian religion and to preach the word of God."[29] Identified with the angel of the sixth seal (Rev 7:2),[30] the pope arrives in a sabbath age (*sabbatum*) signaling evangelization of the entire world, which begins just prior to the arrival of Antichrist.[31] This latter idea, that of an *ecclesia* that engages the forces of Antichrist rather than merely suffering them, is central to Roger Bacon's agenda, though rather less so to Rupescissa, who developed instead the notion of a papal Antichrist. In the traditional understanding of the end times, the role of the Church is largely passive, and it is generally referred to as the subject of oppression and subversion. Many Christians will suffer martyrdom in this period, but the heroes of the apocalyptic drama, such as Enoch and Elijah, come from without. According to Joachim, it is the pope, with other committed Christians, who ultimately triumphs over Antichrist and is regarded as the head of a true universal *ecclesia* at the beginning of the third *status*.[32]

Roger Bacon was actively working on his *Opera* scarcely a decade after the Joachite Gerardo of San Borgo Donnino penned his *Introductorius* in 1254. Just a student at Paris, Gerardo precociously declared that Joachim's prophecies would soon supersede the Bible during the Age of the Holy Spirit. The furor that followed was gratifying to opponents of Joachism and the Franciscan Order, and forced the resignation of the Franciscan Minister General, John of Parma. Reputed to be "a great Joachite" by the sympathetic Salimbene of Parma, John was a convenient target.[33] In the wake of the scandal, the Parisian master Bonaventure of Bagnoreggio ascended to the generalate. Bonaventure was likely friendlier to Joachite sympathizers than rigorist Franciscans made him out to be, but managed to hold the Order together.[34] In the wake of the scandal, however, Joachites of any stripe—including Franciscans—needed to take increased care about how, where, and how much they espoused their views. The initial papal condemnation of the *Introductorius* in 1255 made no mention of Joachim's works, but a regional church council at Arles in 1263 indicated the new danger the Church saw in the writings of Joachim and his supporters. Their wholesale condemnation of the Joachite corpus did not put an end to Joachite speculation, but did begin to push it to the margins.

Bacon, *Scientia*, and Antichrist

While Bacon was something of a rabble-rouser when it came to the adoption of anything related to his *scientia experimentalis*, the influence of Joachim and

Joachism often recedes into the background of his work, and he exhibits some uncharacteristic caution when dealing with Joachim.[35] He did understand Joachism—fairly well as it turns out—but he was hardly fanatical. Bacon likely absorbed Joachism from Robert Grosseteste and Adam Marsh, the most prominent English intellectuals when Bacon was coming of age at Oxford.[36] Moreover, Bacon made his own significant contributions to the discussion of the end times, building on the dramatic recentering of pope and Church as active antagonists of Antichrist.[37] Indeed, apocalyptic fears are at the heart of Bacon's plea to the pope to impose *scientia experimentalis* in the curriculum of Western schools.[38] Not only does Bacon believe that *scientia experimentalis* is necessary to combat Antichrist, but that Antichrist will certainly capitalize on the powers hidden in nature. The battle between *Ecclesia* and Antichrist, then, was not just a spiritual contest. Among the sciences Antichrist will master are those of alchemy, astrology, fascination (a kind of allurement based on changing the essential nature of a person or people, which I will discuss in greater detail below), and words of power—all of which he will use to "pacify" Christendom in the same way Bacon hoped Christendom would conquer its enemies.

It is altogether possible that Bacon was not alone in his fears. The pope had commanded Bacon to send what became the *Opus maius*, *Opus minus*, and *Opus tertium* after Roger had discussed his ideas with Clement while he, the pope, was still Cardinal Guy Folques.[39] Clement's letter does not mention Antichrist, nor does it specify the subject matter of the aforementioned conversation. Still, since Bacon mentions Antichrist in what was likely a prefatory letter to either the *Opus maius* or the *Opus minus*, the specter of Antichrist that lurks throughout Bacon's *Opera* and other works may not have come as a surprise.[40] "All wise men believe that we are not far removed from the times of Antichrist," Bacon wrote to the pope.[41] In his *Opus maius* Bacon advocates the study of, among other apocalyptic authorities, Joachim, as well as the pseudo-Joachite tract, *Expositio Sybillae et Merlini*, but the apocalyptic thought which interests him most is his own.[42] This likely stems from the fact that no prior work joined apocalyptic exegesis so much to the *scientiae* that captivated Bacon. Bacon's apocalyptic outlook is fundamental to understanding his work on experimental science and alchemy, though, like much of his work, Bacon's apocalyptic ideas were taken up sporadically and failed to generate a school of thought. The most notable apocalyptic idea to emerge from Bacon's work, the idea of a *pastor angelicus* (angelic pope) eventually became part of what Richard Emmerson has dubbed "normative" apocalyptic (in opposition to radical apocalyptic such as that of certain Joachite groups and other apocalyptic movements linked to unrest and violence). The success of this one strand of thought, however, has somewhat obscured Bacon's other apocalyptic ideas.[43]

As discussed in the prior chapter, Bacon's elixir would aid Christendom to extend its boundaries by altering the complexion of regions and peoples. This suggests a concern with missiology more generally, which arises in various places in Bacon's works. This concern is hardly restricted to Joachites, but it is central to most expectations of the end times, when the renovation of the world is preceded by great conversions of peoples (most notably Jews). Standard Antichrist stories generally considered this an event exogenous to the Church—the conversions are brought about by heroic prophets such as Enoch and Elijah. In the authentic works of Joachim, however, namely the *Expositio de prophetia ignota* (*Exposition of an Unknown Prophecy*), or the more widely available *Liber de concordia*, Joachim, as expected, sees conversions of the heathens as the work a pope in the final days before Antichrist.[44] Joachite apocalypticism places a premium on the conversion of heathens prior to the battle with Antichrist. Bacon echoes this Joachite commitment in his *Compendium studii*, where the activity of "a most blessed pope," among other things, will bring the "fullness of nations" under the dominion of the Church.[45] The central role played by pope and Church in the Joachite scheme led Bacon to recast widespread missionary activities far beyond the borders of Christendom and consider Christianization of the world on a grand scale. No doubt Bacon's imagination was stimulated by recent events, which suggested that just such an evangelizing mission was already under way. William of Rubruck (whose writings from his mission to the Far East Bacon read) and fellow friars had recently begun to make contact with a number of non-Christian peoples.[46] Bacon's commitment to a kind of final evangelization of peoples based on altering their complexions is echoed in his discussion of geography.

In the *Opus tertium*, in a section on *The Places of the World*, Bacon makes a sevenfold argument on the necessity, and beauty, of geography that echoes his work in the *Opus maius*.[47] Much of his argument concerns itself with the need to create and use accurate maps. At an elementary level, geographical knowledge will prevent missionaries from ending up in the wrong countries or regions, or dying on account of unknown features of climate or other dangers. A richer mastery of geography, however, when married to knowledge of astrology, includes knowledge of the "complexions" of regions and peoples. Thanks to this astrological geography, missionaries will be able to avoid those regions "even more hostile to Latins," or, if forced to go there, might at least bring to bear the proper arguments for conversion.[48] While Bacon makes no overt reference to his alchemical thought in this section of the *Opus tertium*, we can see a clear line of thinking connecting his concern with mission to his later alchemical thought. Given the inherent dangers of missionizing, and the relative few numbers of persons capable of reaching out and converting scarcely known peoples, Bacon's elixir was a meaningful solution. Beaming astrological rays about the world via giant mirrors was, in its way, perhaps more believable than thinking that a small number of friars might be able to reach and convert the heathen of unknown lands. Perhaps if given a century or two, personal evangelization might be possible, but given the looming specter of Antichrist, mass conversion was an urgent question.

Bacon tells us that the "greatest utility (*utilitas maxima*)" of geography arises "on account of the violence of the nations who shall invade the world, namely the Jews enclosed in the mountains of Hyrcania, and Gog and Magog, and the nations enclosed by Alexander behind the Caspian Gates, and on account of Antichrist and his followers."[49] This passage immediately follows Bacon's discussion of the use of geography for missionary work, underscoring the link between his apocalyptic thought and his missiology. The legend that Alexander had enclosed Gog and Magog behind a wall or gates was current in the era of Flavius Josephus and was assimilated into the medieval Antichrist legend.[50] Further, by the twelfth century it had been fused with the notion that the lost tribes of Israel would assist Antichrist, a claim Bacon first made in a parallel passage in the *Opus maius*.[51] The fusion of anti-Jewish sentiment and apocalyptic claims was hardly novel in the medieval period, but what is important in this passage is Bacon's attempt to locate formally both Antichrist and his would-be allies among the current nations and regions—legendary or not—of the known world.[52] Indeed, Bacon is far more wary of the Tatars than of Jews, reflecting, at least, the political realities of the thirteenth century, when the might of the Tatars was well known among European authors.[53] As much as Bacon may have hoped for a successful outcome from Rubruck's mission, he nevertheless gave serious consideration to the idea that the Tatars were the apocalyptic enemy.[54]

On the other hand, Bacon did not specifically identify these foreign groups as Antichrist.[55] He follows Jerome's claim that these peoples would burst out from the East, while Antichrist would emerge from the West. When they met in the middle, having trampled through Christendom, these other groups would declare Antichrist a "god of gods." Bacon believes, however, that understanding the location of these other groups will be key to mounting a defense against Antichrist. "If therefore we shall come to know from what part they will come, we can deduce that Antichrist will come from the opposite part."[56] What we also can deduce, though Bacon does not say it, is that Antichrist will emerge from Christendom. Though Bacon did not develop this idea, there is some tacit evidence that suggests Bacon was open to the Joachite idea that Antichrist would be a false Christian.[57] Nowhere does Bacon repeat the standard trope that Antichrist would be of Jewish parentage and born in Babylon. While an argument *ex silentio* is hardly conclusive, later Joachites would quite clearly identify Antichrist as a false Christian, even a false pope.

If Bacon was unwilling to speculate on Antichrist's identity, he was quite certain that one could know a great deal about Antichrist's allies through the analysis of their complexions. To understand the complexion of a people, one needs to know their region's latitude and longitude, topography, elevation, and climate, as well as their morals and customs, all of which reveal themselves through both cartographical relationships and astrological calculation.[58] Since regional complexion—much like astrological circumstance—affects the composition of an individual complexion, the realities Bacon seeks to comprehend through the science of geography are on a macro scale to the realities provided by alchemy. Bacon's geographical and alchemical studies are part of an organic and unified intellectual program. Marrying fields by means of *scientia experimentalis* allows Bacon to apply alchemical and astrological principles on a scale not considered by most practitioners of these arts.

The influence of apocalypticism on Bacon's *scientia* is felt most keenly, then, in the shape of the curriculum he encourages and the urgency of the need to take up alchemy and related disciplines. While this is an order removed from the kind of relationship of the Christian "perfect body" to the elixir, it does indicate the importance of Bacon's Franciscan ties. Whether Bacon learned Joachism and subsequently joined the Order or became inspired by Joachim after taking his habit is immaterial. The uneasy union between Franciscans and Joachite speculation was, in spite of the controversy it generated within and without the Friars Minor, a defining feature of the Order in this period. Bacon's adoption of Joachite ideas was fused also with the special reverence Franciscans reserved for the pope, whose role as patron and protector of the Order was pronounced in its early history, and whose own fondness for Francis and his Order led to the construction and special role of the Basilica of San Francesco at Assisi.[59]

The famous fresco cycle of Saint Francis that adorns the upper church of San Francesco tells in pictures the well-known story of Francis's life. Centering on key moments that defined Francis and his Order, the Master of the Legend of Saint Francis depicts the pope five times in the twenty-eight images that make up the cycle.[60] Three separate popes—Innocent III, Honorius III, and Gregory IX—are depicted. Among the most famous of the papal episodes is the dream of Innocent III (Figure 1). In this panel, a sleeping Innocent sees a vision of Francis holding up a building that represents the entirety of the Church. The following panel shows Innocent confirming Francis's Rule and, consequently, authorizing the Franciscan movement as a whole. The other panels show continuing papal engagement with Francis, and the panels with Gregory IX are particularly striking. Gregory had been the Order's Cardinal Protector before being elevated to the papacy and had handed down important legislation protecting the nascent Order, but also irking rigorists, as Gregory hoped to make use of the popularity of the Franciscans to combat heresy and spread the faith. In the panel of Gregory's dream, Gregory plays the doubting Thomas, while Francis, the *alter Christus*, displays his side wound. In this panel, Gregory is sleeping to one side, while Francis commands the middle of the frame (Figure 2).

While the artistic program at Assisi had a number apologetic aims, the close connection between Francis, Franciscans, and the papacy is quite pronounced. Pope Gregory, in particular, had been quite close to Francis—he was the Cardinal Protector at Francis's own request—and had even helped write a rule for the Poor Clares.[61] That the Assisi program emerged at the end of the thirteenth century or beginning of the fourteenth, during a time of increasing papal scrutiny of the Order, especially the rigorists, is not surprising.[62] Emphasizing the special relationship between the papacy and the Friars Minor may have been a way of shoring up a deteriorating relationship. Thomas de Wesselow suggests that John of Murrovale, elected Minister General in 1296, may have supervised and overseen the decoration of the Upper Church as part of his anti-Spiritual agenda.[63] The Assisi program both bears witness to and (re)constructs the intertwined relationship between Francis, his Order, and the pope. Roughly thirty years before the completion of the Upper Church frescoes, Roger Bacon had mused on the special role of the papacy and envisioned an even more grand role for it. Bernard McGinn sees in Bacon's references to a future holy pope "the first direct, if tantalizingly

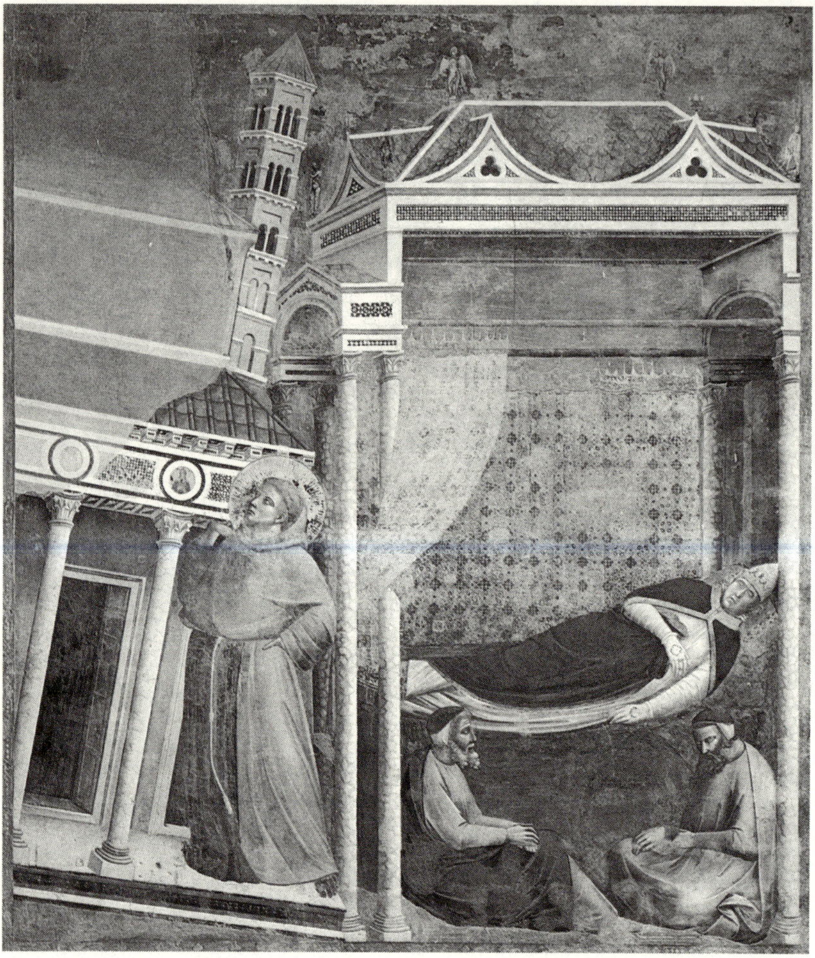

FIGURE 1. The dream of Innocent III. Fresco cycle of Saint Francis, Upper Church of San Francesco, thirteenth century.

brief, witness to the main line of apocalyptic papal hopes that were to crystallize in the fourteenth century."[64] Coming at a time when most popes were canon lawyers (animus toward lawyers is not a modern invention), the idea of a holy pope, a *pastor angelicus*, was a notion that perhaps could only have come from a Franciscan sympathetic to Joachism.

Bacon's notion of the holy pope, however briefly developed, underwent some changes during the course of his life. In his *Opera* composed for

FIGURE 2. Gregory's dream. Fresco cycle of Saint Francis, Upper Church of San Francesco, thirteenth century.

Clement IV, Bacon comes close to implying that his acquaintance, the erstwhile Guy Folques, could be indeed the holy pope himself.[65] Later, in the *Compendium studii philosophiae*, penned soon after Clement IV's death, he took up the issue again, this time outlining a specific apocalyptic context and agenda that draws inspiration from Joachim as well as standard apocalyptic thought. Speaking of the importance of the papacy in the forthcoming apocalyptic confrontation, he writes, "one most blessed pope shall come first, who shall take away all corruption from study, from the church, and other things, and the world shall be restored, and the fullness of nations shall enter the faith."[66]

The central role of the papacy raises an important point about the context of Bacon's alchemy. *Scientia experimentalis,* and the development of all of the tools that are necessary to create the elixir and facilitate the kind of massive conversion Bacon envisions, are not simply the province of interested intellectuals. Rather, it is the Church who is to guide and oversee the development of *scientia experimentalis*, and govern the use of its fruits. Technically, many of the skills of *scientia experimentalis* can be mastered by anyone, but Bacon is not writing simply to spread knowledge or engage in meaningful philosophical dialogue. Alchemy, like all the sciences, is best when practiced by Christians within the hierarchy of the Church for the betterment of Christendom. Knowledge of *scientia experimentalis* is to be regulated.

Bacon's hope for papal support for the educational and philosophical program laid out in the *Opera* did not generate the enthusiasm for which he had hoped. It is in fact possible that he was condemned between 1277 and 1279, though documentary witness via the *Chronicle of the Twenty-Four Generals* comes nearly a century after the events in question. According to the chronicle, "suspect novelties (*novitates suspectas*)" were to blame.[67] Stephen Tempier, Bishop of Paris, issued a series of ringing condemnations in 1270 and 1277 generally aimed at certain Aristotelian tenets, but also including "works of alchemy, necromancy, foretelling or of other superstitions or sorceries."[68] Given Bacon's attempts to rehabilitate the magical as *scientia*, it is easy to imagine how he might have been swept up by—and perhaps even have precipitated—such a condemnation. Whether or not Bacon was formally jailed or charged, around 1278 he had departed Paris for Oxford.[69] Bacon's criticism of the academic establishment during his time at Paris never translated to critique of the Church itself, nor did it afterward. At the same time, Bacon's desire to Christianize alchemy and other sciences does not seem to have captivated the imagination of the Church or the intelligentsia outside of a small group at the papal curia.[70]

Was Bacon's apocalypticism incidental to his conception of the elixir? We know that Christianity itself was not. Bacon's reliance on scriptural and theological models to describe the elixir certainly proves that. But were there apocalyptic ideas that resonated in the same way? One apocalyptic "model" may be Antichrist. Bacon considered Antichrist a master of *scientia experimentalis*, and did so in part to explain how Antichrist would subvert and oppress Christendom.[71] Bacon, like his peers, did not consider Antichrist supernatural. Even the standard legend agreed that Antichrist would rely on the aid of demons in order to accomplish his "miracles." Much as Bacon considered the tools of Christian resistance in the end times to be the product of *scientia*, he likewise employs natural explanations to detail Antichrist's powers. For instance, Bacon argues that Antichrist will use the power of fascination to win over the population.

Fascination—a term with which Bacon is not completely comfortable —is the ability to subdue a populace by changing its complexion. Bacon discusses this idea in a few places, but mentions Antichrist specifically when discussing *The Secret of Secrets.* This text, as mentioned in the prior chapter, contains numerous suggestions of pseudo-Aristotle, among which is, according to Bacon's reading of the text, the utility of bringing nations to heel by fascination. Bacon argues that Alexander used this power, which was responsible for his subjugation of much of the ancient world. This is in part a reasonable—to Bacon—explanation for just how Alexander managed to conquer so much of Asia. (Bacon reasons that Aristotle "handed over the world" to Alexander, as no soldier could match the potency of *scientia.*)[72] At the same time it is also a warning, since Bacon says that Antichrist will employ these same powers, but with much greater effect. We can infer that Antichrist would have some kind of anti-elixir at his disposal, or at least the powers of fascination and *astronomia.* Antichrist is something like a photographic negative of the type of Christian intellectual Bacon wants to create. If Antichrist is able to win over the world, then he will do so using knowledge and power available to human beings. This characterization is dependent on a Joachite interpretation of the role of the Church in the end times. The standard Antichrist legend portrays an almost unstoppable force for evil, which might at best be survived. Joachim's scheme for the end times, however, posited a future where human beings within the Christian Church might offer meaningful resistance to Antichrist. For Bacon, a world where humans play out the cosmic conflict between good and evil requires both sides to be armed with wondrous abilities, and is just the kind of world where an elixir would be capable of possessing the tremendous power ascribed to it.

Vitalis of Furno and Conservative Apocalyptic

In the years of his generalate, Bonaventure was able to contain growing factionalism within the Franciscan Order, in no small part because he actively combated what he saw as laxity among the friars.[73] Shortly after Bonaventure's death, however, the tables turned. Various groups of rigorists among the Order, often the same people who still held to Franciscanized Joachism, became targets of the hierarchy of the Order.[74] And where Joachite sympathies had been cautiously discussed or, like Bacon's, repurposed without much in the way of attribution, persecution had the effect of radicalizing apocalyptic fears and expectations.

In 1283, Peter John Olivi, the French scholastic and theologian who was to be adopted by later Spirituals as well as lay beguines as something of a spiritual father, received his first ecclesiastical censure.[75] Olivi had studied theology at Paris, and his conflict with the Church as well as with members of his own order began in a series of academic disputes. Though he was initially confronted on a range of topics, Olivi later became the chief exponent of a Franciscan ideal of poverty called *usus pauper*, literally "poor use." Olivi argued that all Franciscans had vowed to follow *usus pauper*, which meant generally that they had to restrict their use of goods and live simply except in cases of exigency. Olivi meant for this vow to be flexible, so that brothers could accommodate the needs and commands of the *Ecclesia.* The problem, as his opponents conceived of it, was that without a line in the sand between what is and is not allowed, it was impossible to know whether or not one had violated the rule.[76] Thus the hierarchy dismissed the notion that *usus pauper* was part of the vow.

If Olivi's understanding of *usus pauper* was rejected, however, his attachment to the issue would continue to resonate throughout his work, especially in his Joachite apocalypse commentary, written during a highly charged period in Franciscan history when Franciscan rigorists drew sustained attack from the ecclesial hierarchy. Olivi connected the *usus pauper* controversy to the persecutions suffered by the elect at the hands of the forerunners of Antichrist. He adopted a Joachite division of history, in which a pattern of three *status* is overlaid by seven periods. The lines between *status* and between periods were blurry, and sometimes the transition between one period and the next could span well over a century. Olivi believed himself to be at the transition point between eras (Saint Francis ushered in the beginning of both the sixth period and third *status*) that were simultaneously a time of Antichrist and of spiritual renewal.[77]

At issue for the Order, and later for Vitalis, was Olivi's emphasis on the roles of *usus pauper* and the papacy in the end times. The former issue Burr has summed up succinctly, "Olivi was proposing adherence to his own definition of Franciscan poverty as a litmus test of membership of the elect."[78] According to Olivi, the Franciscan Order—or, rather, the rigorists in the Franciscan Order—took on a predominant role among the protagonists of the apocalyptic conflict. Their spiritual convictions would become synonymous with the surviving *Ecclesia* of the third *status* and seventh period. The problem this posed was that it made the rest of the Church, and the papacy in particular, allies of Antichrist, since they opposed *usus pauper*. Olivi stated that a pseudopope, who he vaguely hinted would be a Franciscan, soon would arise to persecute the elect. His evil nature would be evident through his opposition to Olivi's view of poverty.[79]

Olivi finished his commentary on the Apocalypse just before he died in 1298, but it did not long escape ecclesial censure, nor did his many followers within the Order. In 1299 the Franciscan general chapter banned his writings and threatened excommunication to those who would read, disseminate, or openly discuss his work.[80] Vitalis, who had long frequented Olivi's academic milieu at Paris and Montpellier, but did not share his sympathies for the rigorists, also engaged him in a bitter academic debate in the 1280s.[81] Vitalis's opposition to Olivi did not indicate a lack of commitment to the Franciscan Order, nor, in fact, did Olivi's views on the Apocalypse radically conflict with those of Vitalis. Though Vitalis's commentary is from the same period of time as Olivi's, it harks back to what David Burr has called "the standard Parisian approach to the Apocalypse." Vitalis's commentary shares a common source with both the commentary known as *Vidit Iacob* and the Apocalypse commentary attributed, probably falsely, to the Franciscan Master, Alexander of Hales.[82] There are Joachite elements to the commentary, but radical elements are muted, and it manages to straddle the line between a historical reading of the Apocalypse and the Augustinian proscription against reading the book literally.[83]

Burr notes that a concern with academia is somewhat more pronounced in Vitalis's commentary, and this gives rise to one area of agreement between Olivi and Vitalis: concern over the growing role of philosophy, particularly Aristotelian philosophy, in Christian teachings and practice.[84] Vitalis's commentary speaks of the approach of judgment signaled by the rise of philosophers: "On that day of tribulation . . . .Then shall foolish Plato draw near with his disciples. Then, verily, Aristotle's own arguments will be at hand."[85] In a later passage commenting on the angel sounding the third trumpet, Vitalis seems to be warning of a "great heresiarch, glittering in the night, teaching in the shadows."[86] The implication of the passage is that the heresiarch is learned, and has erred through pride (*superbia*). But how serious could Vitalis have been about Aristotle and Plato, given their omnipresence in scholastic dialogue? Did Vitalis's skepticism in regard to these giants of Western philosophy extend to natural philosophy? Does this rule out Vitalis as the author of *Pro conservanda*, a text that relentlessly cites Aristotle as a chief authority?

Answers to these questions begin with events in Paris roughly a decade before Vitalis arrived, a period in which he found the sources for his Apocalypse commentary. In 1270 and 1277, the Bishop of Paris, Stephen Tempier, condemned the faculty of arts for their dissemination of Aristotelian philosophy.[87] This was not the first time ecclesial authorities had banned Aristotle's works at Paris; this had occurred in 1210, 1215 and 1245. The many condemnations, of course, underscore that the philosophy of Aristotle made significant inroads into the university over the course of the thirteenth century.

Generally speaking, the condemnations took aim at philosophical assertions that led to (or could have led to) conflict with the truths established by scripture and dogma.[88] Among the key persons targeted were Siger of Brabant and Boethius of Dacia, as well as Thomas Aquinas. It appears that by the early thirteenth century, shortly after the founding of the university in 1200, the arts faculty at Paris were including Aristotle (or perhaps Averroës) in their teaching.[89] In 1210 the Archbishop of Sens (under whom the diocese of Paris fell) restricted the teaching of Aristotle's *libri naturali*, that is, his books of natural philosophy. In 1215 the Cardinal and legate Robert of Courçon reaffirmed this decree, adding to the list Aristotle's *Metaphysics*.[90] The proclamations were less a general condemnation of Aristotle than an attempt to corral the putatively heterodox claims of Amalric of Bene and David of Dinant.[91] Private study of Aristotle remained licit, and the prohibition did not extend to the faculty of theology.

In 1231, Pope Gregory IX reaffirmed the prohibition against the teaching of Aristotle, although in this case he ordered a commission to expurgate the Stagirite's works so that they might be used. The commission never concluded its work, however. In the same year, the pope also issued a letter releasing the public study of Aristotle for a period of seven years, which was renewed again in 1237. Soon after, masters of the arts faculty, possibly beginning with Roger Bacon, began lecturing on Aristotle's natural philosophy.[92] By 1255, knowledge of Aristotle was required for candidates in arts, and, as John Wippel stated succinctly, "the faculty of arts had now become a philosophy faculty."[93] This about-face opened the door, however, to radical Aristotelian approaches that did not sit well even with other Aristotelian scholars.[94] Siger of Brabant in particular seems to have been the primary target of thirteen articles of the 1270 condemnations and also drew criticism from Bonaventure and Thomas Aquinas, whose works would be targeted in 1277. In addition to Siger, Boethius of Dacia was targeted in 1277, though neither name was mentioned in the text.[95]

Among the opponents of the radical Aristotelian movement were two Franciscans, Bonaventure and John Pecham (d. 1292). Pecham, for his part, was a foe of Thomas Aquinas, and later became Archbishop of Canterbury (1279).[96] Another significant voice of opposition was Henry of Ghent, whose philosophical works Vitalis of Furno followed closely (or copied outright).[97] These scholars objected to the notion allegedly championed by Siger that philosophy could arrive at a second truth in addition to truth provided by revelation, though Siger's works show no evidence that he actually held this belief.[98] Siger did have an Aristotelian view of God's role in the world, however. Siger held that God was the first cause, but did not view God as directly responsible for creation. God intervened directly only once. Everything else that followed (including the creation) was a necessary result of the first and only divine act.[99] It is difficult to reconcile Siger with the world at his doorstep, one awash in saints, miracles, and manifest interventions of God, and indeed it is likely that Siger was not as attached to these views as the condemnations might suggest.[100]

While many censured propositions would have required some years of philosophical training in order to understand their relevance to Christianity, a clear and overriding concern of the condemnations is any denigration of either scripture or the faith.[101] Underscoring this is that the propositions entail possible errors as much as real ones. Roland Hissette was able to determine the exact author targeted in 151 of the propositions, but in nearly every case the actual thought of the author was misrepresented or taken to such an extreme as to become heterodox.[102] While sanctions were real for those explicitly targeted, one should not assume that Tempier and his commission of doctors misrepresented or misconstrued hundreds of philosophical statements. It seems that the condemnations were meant as much to obtund future speculation as they were to expurgate current philosophical conceptions leveled against the Christian faith. As J. Thijssen has stated, "there were no medieval scholars who *opposed* philosophical conclusions and statements of Christian doctrine, nor did they defend an untenable theory of double truth."[103] Even Roger Bacon, who championed the study of philosophy, considered it sinful to believe that theological truth should flow from, or be in any way subordinate to, secular philosophy.[104]

These concerns seem to be at the root of Vitalis's condemnation of Plato and Aristotle. A conservative theologian and political climber, Vitalis was echoing the orthodox establishment's claims about the dangers of Aristotelianism. While the tentative date of 1292–1307 for Vitalis's commentary might seem a rather late response to the 1277 condemnations, two facts argue for its topicality. First, Vitalis is relying on older commentaries, including one attributed to the team of Hugh of St. Cher written perhaps in the 1260s.[105] Second, the condemnations were not a sudden event—the controversy over Aristotelian teaching was brewing long before the 1270 and 1277 condemnations and lingered well after. The concerns over secular philosophy were part and parcel of the Parisian theological community throughout much of the thirteenth century, and Vitalis is wholly consistent with conservative hardliners as well as prominent scholastics of his own order.

Skepticism over Plato and Aristotle was limited to competing cosmologies and double truths—fear of corruption of the faith. Vitalis makes this plain enough in his Apocalypse commentary when he later describes physics as "the crown of eternal salvation," logic as "the expulsion of devilish deceit," and ethics, "the mirror of all knowledge."[106] Just about every academic subject one could name receives a similar paean, including natural philosophy (*naturalis sciencia*), geometry, astrology, and mathematics.[107] Vitalis, like Stephen Tempier before him, was not critiquing the liberal arts or philosophy in a general way, however negative his condemnation of Aristotle and Plato may have sounded. Rather, he was condemning the danger that awaited those who approached Aristotle too slavishly or unwarily. Aristotle and Plato were useful insofar as they did not contradict revealed Christian truth. Other knowledge, discovered and discussed under the canopy of Christian doctrine, was to be celebrated.

In this sense, Vitalis's apocalyptic view is consistent with the kind of philosophical discussions found in Chapter 1, where Aristotelian discussion of optics and physics emerged from a close reading of Genesis. Provided scripture and other Christian truths remained constant and unquestioned, antagonism toward the sciences was unnecessary. This is reflected in Tempier's condemnations themselves. Looking at the condemnations that touched on alchemy, eleven stand out. Of these selected propositions, we can categorize them generally around the principles of free will and of the nature of the elements. Most, however, deal with the nature of the heavens.[108] As discussed in the prior chapter, Aristotle's *Meteorology* and commentaries upon it bore directly on alchemical pursuits, owing both to the influence of the heavens on their earthly counterparts and, more directly, to the idea that metals were heavenly emanations congealed within the earth.[109]

Rather than contradicting alchemical ideas, some of the condemnations indirectly endorsed them. Condemnations of propositions 111 and 112 ("That the elements have been made in a previous generation from chaos; but they are eternal." "That the elements are eternal.") reinforce the mutability of the elements. Likewise, the wording of the condemnation of proposition 101 ("That God could not have made prime matter without the mediation of a celestial body.") asserts that prime matter existed. Rather than condemning any kind of speculation on the natural world, the condemnations simply assert that such speculation cannot be correct if it controverts the revealed truth of Christianity.

If we take Vitalis's opposition to Plato and Aristotle seriously—and there is every reason we should (he could have, after all, chosen to plagiarize a different apocalyptic treatise)—it sheds additional light on the alchemy attributed to him. First, there is nothing in his apocalyptic condemnation of these two Greek philosophers that suggests an opposition generally to worldly knowledge. Medicine is a perfectly appropriate science for the Christian to pursue. Moreover, the genre of the *materia medica* would have allowed Vitalis to sidestep the kind of cosmological questions and assumptions that might have led to scrutiny. Since *Pro conservanda* is not a treatise, the simple, often under-theorized descriptions are adequate. The Aristotle cited so frequently in *Pro conservanda* is the kind of Aristotle Vitalis the apocalypticist is comfortable with—an Aristotle that does not provoke the kind of dangerous theoretical questions presumed a generation prior by Bishop Tempier. Instead, the Aristotle of *Pro conservanda* is simply an authoritative and astute guide to the presence of various secrets in the natural world, none of which require adherence to a non-Christian cosmological theory or, really, a non-Christian belief of any kind.

Finally, Vitalis's stance on Aristotelian and Platonic philosophy, that it is licit if pursued within a Christian framework and subservient to revealed truth—places him on a continuum with Roger Bacon and John of Rupescissa, as will be shown forthwith. Vitalis simply does not pursue the union of natural philosophy and Christian thought to the same extremes. While it is possible to see Vitalis as championing naturalism versus the supernaturalism or hypernaturalism of religious alchemy, it is more useful and perhaps more apt to say that he shares with his brethren some common assumptions about the role of Christian truth in the pursuit of natural philosophy and alchemy, however much his work may differ in the end. We can take these arguments only so far, however, given the limits of what we know about the authorship of the texts ascribed to Vitalis, but the conservative apocalypticism of the Franciscan Cardinal fits the conservative approach to alchemical theory and the elixir within *Pro conservanda*.

The Olivian Legacy: John of Rupescissa

Vitalis of Furno came to an unfortunate end. His political rise coincided with increasing papal scrutiny of the Franciscan Order. Early in Vitalis's academic career, Pope Nicholas III had issued *Exiit qui seminat*, a papal bull that guaranteed the ability of Franciscans to carry out the tasks appointed to them by the papacy without compromising their vow of poverty. This endorsed a legal fiction earlier created by Gregory IX, whereby the pope owned and had dominion over all goods and property used by the Franciscan Order. The Franciscans could therefore own nothing, but would still have access to the materials necessary for preaching, scholarship, inquisitorial practice, and care of souls. As importantly, the bull also characterized poverty as a Christian ideal, and a virtue taught and lived by Christ himself.

Both of these ideas came under attack. The idea of papal dominion of goods found opponents from within and without the Order. Rigorist Franciscans were dismayed by the Order's access to what they considered luxuries, and in the 1280s Olivi began his intellectual campaign to rein in excess. Olivi considered all Franciscans to have taken a vow of evangelical poverty, which meant both the renunciation of dominion over goods and property and the observation of *usus pauper* as discussed above. To reiterate the key element of this definition, poor use did not mean that Franciscans might never ride a horse, for instance, but would do so only in cases of necessity. Poor use was a kind of middle ground between absolute poverty and the needs engendered by the duties of an active international religious order.[110] Olivi championed this idea in his Apocalypse commentary, in addition to other places, leading to an all too easy transposition of philosophical opponents to apocalyptic enemies. Though he does not seem to have targeted any of his contemporaries, his followers would not be so hesitant.[111]

The dispute over the absolute poverty of Christ was another key issue in ecclesiastical suspicion of the friars. Not only were rigorists accused of celebrating themselves as more spiritual than their brethren because of their poverty, but the Franciscan commitment to poverty aroused the ire of other religious orders. The Dominicans in particular were rankled by Franciscan claims to be specially following in the footsteps of Christ, and by 1320 they actively—and successfully—attacked the Franciscan opinion that neither Christ nor his apostles owned anything. This idea was present in *Exiit qui seminat*, and was not at issue among the Franciscans—both rigorists and conventuals defended this idea.[112] Dominicans up to this point had been concerned chiefly with rooting out heterodox Spirituals, but the current pope, John XXII, was so vexed by the Franciscan problem that he was eager to put it to a swift end regardless of the consequences. In 1322, he issued the first of three bulls in a span of two years that eviscerated not only the rigorist wing of the Order, but much of Franciscanism in general. The first bull, *Quia nonnunquam* (1322), overturned injunctions that forbade discussion or critique of *Exiit qui seminat*. Two versions of *Ad conditorem* (1322/23), which rejected papal dominion over Franciscan goods (in effect making Franciscans possessors of the properties they inhabited and the goods they used), followed. Finally, *Cum inter nonnullos* (1323) asserted that Christ and the apostles had dominion over goods, and rejected the doctrine of absolute poverty. Moreover, anyone who claimed otherwise was to be regarded as a heretic.

The fallout was swift and brutal. The erstwhile papal ally Vitalis, who was aghast at the pope's desire to solve the Spiritual problem by effectively dismantling the legal and ideological foundation of the Order, was shouted down in consistory in 1322, and died soon after. Michael of Cesena, the Minister General of the Franciscans who was unsympathetic to the Spirituals, was nonetheless compelled to oppose the pope and took refuge at the court of Louis of Bavaria. The Franciscan Order was fractured, though not destroyed, and many followers of Olivi believed that they had found their pseudopope.

John of Rupescissa has been held up as an exemplar of radical Joachite Franciscanism in the era following 1323, but it is important to state that such claims have more to do with Rupescissa's reliance on Olivian Joachism than with any loyalty to the Spiritual program.[113] John rejected two important characteristics of the Spiritual Franciscans and Fraticelli: the adherence to the Olivian conception of Franciscan poverty (*usus pauper*); and, the belief that Pope John XXII was a persecutor of the Franciscan Order (or even Antichrist).

In the Olivian understanding of the third *status*, a time of terrible tribulation by the Antichrist and antichrists, the Christian faith will at the same time be renewed by a group of *viri evangelici* (evangelical men).[114] As conceived by the early Spirituals, this term generally referred to holy men who lived according to Saint Francis's testament, that is, according to the tenets of evangelical poverty.[115] These evangelical men were possibly Franciscans, but this distinction is not always made, nor does it reflect the many variations of this prophecy. For Olivi, and for the Spirituals and Fraticelli who later championed his apocalyptic scheme, as far as the Church of the end times was concerned, *usus pauper* became one of its defining features. Olivi and his followers argued that the Church itself will forego possessions and dominion and appear again as it did in the apostolic era.[116] As David Burr succinctly noted, the Church of the third *status* is "Franciscanized." So too are its members, the *viri evangelici.*[117]

Following Olivi, John of Rupescissa envisaged the *viri evangelici* as a remnant of the Friars Minor, or at least like them, but he did not identify them as a current Franciscan party—certainly not conventual friars, but also not followers of Michael of Cesena, French Spirituals, or Italian Fraticelli.[118] For while Rupescissa was keen to emphasize that this remnant would be persecuted, one of their key characteristics is that they denied neither papal authority nor the unity of the Church.[119] John sympathized with the desire to reform the Church spiritually, and condemned the excesses of the Church and some of his brethren, but he affirmed the validity of John XXII's regime (even at one point calling him an instrument of divine providence!) as well as that of his successors.[120] (Indeed, it was to the papal curia itself that John pled his case.)

In the eleventh tract of the *Liber ostensor*, in which John predicts the fate of various orders, he seems to take particular aim at the Spiritual Franciscans.[121] Citing the examples of Noah and Lot, who John argues did not withdraw from evil men until their doom was at hand, he states that "no evangelical poor man ought to divide himself from the unity and obedience of the prelates, especially the pope and cardinals, but ought to be able to beware the contamination of all sins."[122] Thus, "the evangelical poor, the professors and defenders of the *highest poverty* are those who never withdrew from obedience and the unity of the pope and cardinals nor from Pope John XXII" (emphasis mine).[123] This passage suggests that John's Franciscan identity is predicated more on obedience to clerical authority than on a (rejected) doctrine of poverty. This is not to suggest a defense of profligacy among the friars. John takes special aim at the Dominican Order as a group of "mammonist heretics (*mammonistas hereticos*)," and elsewhere castigates the lax (*laxi*) among his own Order.[124] John's view of poverty, however, recalls, perhaps self-consciously, a pre-Bonaventuran expression of its devotional merit rather than its doctrinal definition.[125] John extols poverty, but terms such as *usus* and dominion, the key terms of the poverty dispute, do not arise in his discussion. No Franciscan would have been unaware of the import of such terms—and the danger they carried—and John's hortatory discussion must have omitted them purposely. However keenly John felt his own persecution at the hands of the Church and his fellow friars, he was living in a world after *Cum inter nonnullos*, and refused to resurrect the divisive argument over poverty that one might expect to accompany his Olivian prophetic scheme.

John further divides himself from the Spirituals later in the *Liber ostensor*, when he attacks noted Spirituals Angelo Clareno (d. 1337), Ubertino da Casale (d. 1329), and Philip of Majorca (d. 1342).[126] He reproves them generally for dividing the Order, but each is treated specially. Angelo, he says, lived "up to this day without a head (a minister general) and without obedience and without correction."[127] Ubertino was guilty of "rending (*scindens*)" the friars and living against the rule (*contra Regulam*).[128] John derides Philip for his arrogance, for compiling his own Rule as if able to practice poverty more perfectly than Francis himself.[129] Unfortunately, but not unsurprisingly, John offers no substantive commentary on either John of Valle or Gentile of Spoleto, from whom the nascent Observant movement sprang.[130]

Some of John's conservatism on this matter can be explained by Robert Lerner's theory that John learned his Joachism not from a clandestine Spiritual source, but from Guirel Ot, the Provincial of the Order in Toulouse (and later Minister General of the Order).[131] In his commitment to a devotional definition of poverty and in his unwavering obedience to the papacy in spite of their findings against him, Rupescissa arguably held closer to Saint Francis's tenets than did either the Spirituals or the Michaelists (though the existence of a papal Antichrist, as the Spirituals and Michaelists figured John XXII to be, would seem to render this distinction moot).[132]

Though John tends to foreground obedience and unity over poverty, he did understand the *viri evangelici* to be poor. In fact, their presumed poverty is the reason he devotes time in *De consideratione* to describing methods for producing the elixir without resort to expensive materials.[133] Just as much as Bacon did, John envisioned the elixir to be a weapon against Antichrist. The ability to conduct distillation without expensive equipment would allow the evangelical men to actively oppose Antichrist (not merely endure him), while remaining true to their spiritual calling. John's 1350 alchemical tract, the *Liber lucis*, provides a means to enrich the Church as a corporate body, "in order to solve the grave destitution and future poverty of the holy and elect people of God" through the production of alchemical gold.[134] There is a parallel here to Roger Bacon's work, in that John believes it will be necessary to ensure that the Church has the ability to take an active role in the apocalyptic drama.[135]

In any case, the *viri evangelici* are Franciscan in character, if not in habit or affiliation.[136] The chief characteristic of these men is complete evangelical poverty, as imagined by Saint Francis and encouraged in his *Rule* and *Testament*. That is to say, John self-consciously harks back to a Franciscan poverty untroubled (and undefined) by the Olivian doctrine of *usus pauper* and its subsequent papal condemnation, or by the realities that begged such a definition in the first place.[137] Their special sanctity and abnegation of worldly temptation made them worthy converters of the people and a new coterie of martyrs upon whose blood the *renovatio* of the Church will be based.[138] No special distinction is made between the evangelical men of Rupescissa's time and those of the future, because no such distinction was necessary. Spiritual truth and virtue were eternal. Much as Saint Francis could be an *alter Christus* some twelve hundred years after Jesus's death on the cross, so too will each of these new saints be an *alter Franciscus*.[139] During the advent of Antichrist, however, John obviously does not envision a Franciscanized Church. As mentioned above, John issued an alchemical treatise on the Philosopher's Stone precisely as a means to restore the Church's wealth.[140] Like Roger Bacon, John has in mind an active, even militant Church in need of both coin and tools to combat Antichrist. This implies that John understands that, while the evangelical men will be poor and Franciscan in character, they will not practice absolute poverty as defined by Olivi, nor will Christian resistance to Antichrist be merely passive.[141]

John's treatise on the elixir, the *De consideratione*, is an attempt to arm evangelical men with the resources to combat the devil and his minions, to provide a substance "for protection in every time of war and tribulation, and especially in the time of Antichrist."[142] John's elixir, the quintessence, is an impressive substance, though it is not nearly as impressive in its reach as Bacon's elixir. If Bacon's elixir was the secret to ushering in the Christian triumph at the end of days, John's elixir was more of a tool to be used during Antichrist's oppressive encroachment. The healing properties of the quintessence are its most obvious advantage, as it can heal major battle wounds and revive the near-dead.[143] Like Bacon, however, John understands that the effect of the elixir on the body impacts the mind as well. For instance, the quintessence, when made with peony and angelicum, imbues its drinker with courage. Since the quintessence is principally distilled wine, the effect seems quite reasonable, though John's advice to hand it over to the armies of Christendom might be somewhat counterproductive.[144] Other effects are subtler, such as the ability to drive away demons, confusion, and despair, implying that the quintessence would be effective against the otherwise convincing preaching and pseudo-miracles of Antichrist. John's apocalypticism is more closely aligned with his alchemical undertaking, though unlike Bacon, he does not envisage the war against Antichrist being won by natural philosophy. John instead has described a medicine capable of aiding the faithful by giving them strength and comfort.

* * *

It is not clear that we can reliably or productively separate the impetus for the alchemical pursuit from the alchemical product at hand. It seems foolish to suggest that Bacon or Rupescissa did not envision the needs of the end times before suggesting an appropriate remedy. Their adoption of particular intellectual traditions and genealogies was likely governed as much by their goals as by the persuasiveness of the intellectual traditions themselves. And, chronologically speaking, both of them wrote of the end times before expounding their alchemical theories. Even in authors where the link between apocalypticism and alchemy is weak, such as in Vitalis's works, disentangling alchemical thought from apocalyptic thought is highly problematic. In general, it presupposes that differences between medieval genres—alchemical treatise, apocalypse commentary, visionary manifesto—represent mental boundaries that preserve one discourse from the influence of another. Such a presumption is wrong on its face, given the fact that Bacon's and Rupescissa's alchemical musings are littered with apocalyptic references. Even in the case of Vitalis, a rationale for a moderated study of natural philosophy is presented in an apocalypse commentary. It is likely that modern scholars are more to blame for this artificial division. We have canalized ourselves into the history of religion or the history of science, or other such subfields that tacitly presume that disciplinary boundaries can be retrojected, even if consciously we are all well aware this is unlikely.

To return to where I began this chapter, Leah DeVun has made a convincing argument, with which I agree, that Rupescissa's metaphorical discourse did not suggest that Christianity was simply an impetus, nor even an analogical vehicle, to communicate alchemical practices.[145] Rather, the use of Christian metaphor was foundational for how Rupescissa understood alchemy, and alchemy shaped his understanding of Christianity. DeVun has argued that Rupescissa's thorough immersion in liturgical practice, theology, and apocalyptic speculation undoubtedly guided the way in which he expressed alchemical ideals, such as the quintessence being human heaven.[146] Alchemy and apocalypticism are linked by more than language, however. Each, in the formulations of Bacon and Rupescissa, manifested spiritual realities on the physical plane. The advent of the end of days took the cosmic drama of good versus evil and located it within the world. Bacon's elixir did much the same for good. John's elixir is an even more stark example of this shared aspiration, as John believed one could distill and manipulate the literal stuff of heaven—the awesome miracle-making power of God.

The union of apocalypticism and alchemy spoke to a deep-seated need to realize Christian truth in a physical way. This is not to psychologize the authors, but to link these discourses to a host of devotional practices around the bodies of saints, the Eucharist, and other physical manifestations of holiness. Therefore, the alchemically enhanced body and the end times offered the possibility of a kind of heightened experience of embodied spirituality that potentially transcended more widespread tools of material devotion. Bacon's and Rupescissa's union of apocalypse and alchemy placed the practitioner in a proactive role that focused both on saving one's own soul and on saving Christendom. The end times also presented a danger to the faithful. Antichrist would lead astray those of weak will, and his persecution of the faithful would be unrelenting. The psychic effects of the elixir, however, would ward off this danger. It could act as a kind of guarantor of one's enduring Christianity in spite of the trials that awaited all in the end of days. The elixir's ability to make or maintain Christianity, however, also suggests that the production of the elixir required something more than a manual of instruction. It required the world to work subjunctively. That is to say that the world needed to be as it *should be*, not always what it was, for distilling heaven required the assent or even the participation of the divine. It is to this particular ingredient—the means by which it was pursued, and how it should influence our reading of these alchemies—to which this book turns next.

CHAPTER 4

A Subjunctive Science

Most of this book has dealt with Franciscan intellectual life—the religious and scientific discourses that were intertwined in discussions about the elixir. In this chapter we turn to practice. In what follows I answer the question of how we characterize alchemical operations as described by the various authors we have encountered.[1] In prior chapters my method was to examine religious discourses that enmeshed themselves in alchemical pursuits. In this chapter I turn more toward religious praxis, namely ritual practice. Of course, there is still a discourse to be dealt with here, namely the discourse(s) about ritual. Still, these discourses can be informative about praxis, provided that self-definition does not obtund other interpretations. As I laid out in the introduction of this work, scholars such as Eliade and Jung have considered alchemy ritually, but did so in a way that considered the material practices of alchemists as accretions and subversions to what should be entirely spiritual pursuits. My approach is wholly different. First, I do not consider alchemy necessarily to be a ritual practice. Yet, just as outside discourses such as those on apocalypticism influence alchemical discourse, so too might exogenous practices inflect alchemical praxis. Second, I privilege the materiality of alchemy. The physical operations and products of the alchemical pursuit were precisely what gave them their meaning.

Given the rather weighty intellectual achievements of the Franciscan Order as well as the various intellectual and legal disputes that engulfed them in the Middle Ages, it is easy to lose track of the fact that Franciscan daily life was punctuated not by Joachite speculation or philosophical inquiry, but by worship. Franciscan life was more flexible than that of monks, but the Friars Minor were expected to keep up a regular liturgical schedule based in part on the monastic hours. If we cannot imagine Roger Bacon, Vitalis of

Furno, and John of Rupescissa participating in daily religious rituals, then we have not fully imagined them.

Showing a relationship between praxis and idea is not as straightforward as uncovering genealogical and discursive connections among thinkers, where the evidence of connection more or less makes itself known through common ideas or discursive breaks and irruptions. Practice has its own genealogy, albeit one that leaves only traces in the written record. Pamela Smith has noted the fact that early modern alchemical practice was not discursive knowledge, but rather a knowledge of the body, passed on via observation and imitation, as the full range of praxis was too complicated to be consigned to written tradition.[2] At the same time, there is always an improvisational possibility in practice, otherwise it would be incapable of adaptation. It would be foolish to suggest that nonnarrative practices in a chemical laboratory today have nothing to do with *chymical* practices in early modernity. Since the survival and transmission of technique does not rely only on the written medium, I focus in this chapter on how these Franciscans alchemists theorized their practice. How did they understand the way in which one made the elixir? To this should be added, how did they reconcile themselves to the fact that none of them could have made the elixir they endorsed?

Answering these questions is not a straightforward affair. I will begin then with what Bacon and Rupescissa claimed they were *not* doing, namely magic, and Rupescissa's equivocation about whether or not he was producing miracles, that is, whether his alchemy amounted to theurgy. Magic and theurgy were associated with alchemy in the Hermetic corpus of late antiquity, and though I have generally hewed to the medieval context in this book, some discussion of the Hermetica is required in order to distinguish the difference between the medieval approaches to the elixir and the miracle-working of late antiquity. Next, I turn to an examination of Franciscan ritual practice as a means of understanding how Bacon and Rupescissa conceive of their practices. Ritual does not—and should not—necessarily imply a kind of praxis that engages the supernatural or is any way irrational. Rituals may seek to invoke the divine, but they are entirely the product of human behavior. Therefore it is imperative that we approach rituals in terms of how they shape human experience. The Eucharist, baptism, feast days, and last rites punctuated the course and rhythm of medieval life—even more so for members of a religious order. Again, I do not suggest that alchemy is, in itself, a ritual. Rather, the ritual life of Franciscans provided a means for interpreting the contingent and uncertain results of alchemical practice. Ritual ordered (and still orders) humans in relationship to one another as well as to God and the cosmos. At their heart, rituals are regulators of social interaction by producing the world *as it should be*, not necessarily as it is.[3] Adopting the idea of the subjunctive allows Bacon and John to straddle the making of a substance that is both natural and not—the pursuit of a human endeavor, yet one that begs for the endorsement of God.

Magic and Miracles

However much Bacon's and Rupescissa's alchemies were intertwined with Christianity, they were nevertheless problematic. The lengths to which Bacon and Rupescissa go to assert their orthodoxy and the licit nature of their work clearly indicates that readers of their texts might very well read their treatises with suspicion. Franciscan statutes show that assumption to be correct. The Order condemned alchemy repeatedly; the Chapters of 1260, 1279, 1292, 1313, 1316, and 1337 offer various elaborations, though penalties were more severe in the late 1290s than in, say, 1313 or 1316.[4] The definitions between 1295 and 1318, however, do have some common elements, particularly a linking of alchemy with occult or sorcerous practice.[5] Brothers are prohibited from

> works of alchemy, necromancy, foretelling or of other superstitions or sorceries or any other operations of suspect teachings or arts that are not taught in public or have been condemned by the church, generally any deceitful or hateful operations such as invocations of demons, and incantations of people and things.[6]

Franciscans are also prohibited from teaching said arts, as well as having, reading, making, or commissioning books or pamphlets on these nefarious practices.[7] To outsiders, alchemical processes were not easily distinguishable from magic or sorcery. Alchemy was lumped together with other practices believed so dangerous that practitioners were threatened with excommunication, prison, and, in the case of dead suspects, the exhumation of their corpses (and presumed re-interment [or not] outside hallowed ground).[8]

Bacon, as noted already, had some trouble with the term "magic," sometimes using it as a term of opprobrium and yet also describing his educational program as the study of "magical sciences (*scientiis magicis*)."[9] In spite of the somewhat casual use of the term in the *Opus maius*, Bacon argues, largely on semantic grounds, that many techniques may appear to be magical, but have been misconstrued as such. One of the places where this is evident is in his discussion of astrology, where he takes issue with Latin scholars who conflate the terms *mathesis* and *matesis*. After grumbling that the small-minded grammarians (*glommerelli*) are too ignorant of Greek to understand the difference, he echoes the distinctions made by Michael Scot, John of Salisbury, and Hugh of St. Victor in regard to these terms.[10] *Mathematica*, aspirated in the middle where it is spelled with an "h," denotes mathematics, or mathematical astrology, whereas *matesi*, unaspirated and shortened in the middle, refers to diviners or soothsayers. Bacon, like Hugh, saw *matesi* derived from the Greek *manteia* (or, rather, *mancia* as he transcribes it), namely, oracle or prophecy, and believed that grammarians had mixed up the two terms. Legitimate philosophy was the domain of *mathematicians*, oracles and divination of *matematicians*. Though technically referring to diviners, Bacon expands the definition of *matematicians* (sans "h") to describe magicians in a more general sense. Likewise, the term *mathematician* becomes synonymous with philosopher.

Bacon's general concern is with the idiocy of magicians who fail to understand the natural causes of many allegedly magical effects. This is apparently more dangerous than demonic magic, which Bacon disregards as a lesser evil. "True philosophers have never worried about invocation of demons, but only insane and cursed magicians do so."[11] Bacon contends that left to their own devices, without demonic aid, magicians are prone to even crazier flights of fancy than if under the influence of malicious spirits.[12] Bacon, was, of course, no stranger to demonic magic. In his critique of educated clerical and secular men who resort to necromancy, Bacon lists a number of necromantic manuals—often different titles in different works. This suggests that Bacon's opposition to necromancy was largely rhetorical and that he well understood just the sort of activities in which learned sorcerers involved themselves.[13] He devotes more attention, however, to the manner in which false magicians proceed, namely, the use of incantations, sigils, charms, and the like, which he groups together with sleight of hand and trickery.[14] In the *Opus tertium*, he allows for the fact that demons likely do the bidding of magicians who summon them, but the argument in the *Tractatus brevis* takes a separate tack and treats most sorcerous activity as "purely magical (*purissime magica*)," that is, sinful, activity, but not demonic.[15] Bacon characterizes it more generally as error, and his critique here is targeted at the ignorant as much as the malicious. Apart from necromancy, there is essentially no magical activity that does not correspond to some aspect of *scientia experimentalis*. Astrology, as discussed, was perfectly licit when freed from determinism. Bacon also considered *fascination* the result of scientific knowledge. More examples of Bacon's semantic distinction between science and magic will present themselves shortly, but what should be clear is that Bacon's objection to magic is framed by how close he comes to engaging in it. Bacon never ties alchemy explicitly to magic, possibly because he engaged in or showed enthusiasm for much more suspect practices.

The issue of magic is perhaps less complex for Rupescissa, though he strays much closer to the edge of heterodoxy. John knew that the power of his alchemical creation, and the method by which he derived it, would draw suspicion. Hence he asserts that what he was putting down in writing was not a spell. The creation of the *aqua ardens* was "without the sin of magic," but John readily admits that witnesses would believe it to be so.[16] One presumes it is because of the awe-inspiring nature of the elixir itself rather than any process or activity John engages in. For instance, he claims that God provided knowledge to circumvent some of the expenses and time associated with the process of creating the quintessence, but most of these are quite mundane. One means is the less than miraculous burying of the distillate-to-be in a pile of "the best" horse dung in order to heat it.[17] This seems a fairly straightforward affair. Yet, John himself more than once refers to the creation of the quintessence by these means as miraculous.[18]

Where Bacon is on guard about being accused of magic, John comes very close to (and occasionally embraces) the idea that he is alchemically producing a miracle. Again, semantics come into play here. John writes from the perspective of a person in a sacred universe. Hence, John's understanding of miracles is quite expansive. On the one hand, he includes a number of what in a different context might be considered natural substances. The herb hierapicra, for instance, belongs among the "miracles created by God in this world," owing to its medicinal applications.[19] On the other hand, John frequently wades into supernatural territory, telling his readers that with his instruction "you will do the work of God rightly, and through His power make miracles upon the earth."[20] The quintessence, argues John, is God's tool for working miracles.[21] He resorts to an awkward metaphor of a carpenter who uses an adze for making boxes. The adze is simply a tool, but it is not, in and of itself, integral to the carpenter's ability to make a box. If the carpenter did not have the adze, he could still make boxes, albeit in a different way. Accordingly, God uses the quintessence as a tool, by bestowing celestial bodies with quintessence in order to achieve perfectly His desired aims.[22] The quintessence is not *necessary* to miracles—a point John stresses in his struggle to cling to orthodoxy—but his argument harks back to scholastic discussion on the heavens. Here John is referring to the concept that the celestial region affects the terrestrial. The idea of "influences (*influentias*)" proceeding from heaven to earth was used to describe all the effects of heaven that could not be attributed to either light or motion from the celestial region. John adheres to the specific idea that the influences of heaven are responsible for the production of metals.[23] Hence, gold, which is imbued with the power of the sun, the preeminent celestial body, is the most potent healing metal.[24] Still, much as Bacon guards his astrological claims by stating that astrological conditioning is subject to God's will, so too does John consider that the miracle-making power of the quintessence is subject to divine prerogative.

A Hermetic Legacy?

The ties between magic or miracle-working (theurgy) and alchemy are strongly associated in late-antique Hermeticism, a religio-philosophical tradition that emerged in Roman Egypt. Many of the texts (philosophical or technical) were ascribed to Hermes the Thrice Great (Trismegistos), but are often categorized according their general combination of Aristotelian empirical philosophy and the notion that all objects and phenomena are connected through *sympatheia*, the idea that divine energies connected the cosmic to the terrestrial in a hierarchy of being.[25] Zosimos the Panopolitan is most often associated with Hermetic alchemy and was one of alchemy's pioneering figures in his own right. In addition to developing technical instructions (including a well-known diagram of distillation) for the manipulation of matter, Zosimos argued that comprehension of the properties of matter and material relation was necessary for the liberation of the soul.[26] While the salvific aspect of Zosimos's alchemy is engaging, it is too theologically dissimilar from the thought of Bacon and Rupescissa to be of use in the present discussion. Of course some Hermetic texts were translated in the Middle Ages, including the widely known *Emerald Tablet*, but the context in which these texts emerged—the religious and cultural melting pot of Roman Egypt—was likely lost on the vast majority of even educated readers in the medieval West. This need not mean, however, that the texts were not read, adapted, or viewed as anything less than consonant with current practice.

The Hermetic tradition encompassed a wide range of works dating from the century before Christ to roughly four centuries after. Among the characteristics that bind this variety of texts together is their ascription to Hermes Trismegistos and a number of associated characters, many of whom were semi-divine. Hermes Trismegistos was the creation of Hellenic Egypt, something both more and less than the fusion of the Egyptian deity Thoth and the Greek deity Hermes as a kind of divine man, or a cipher for the divine. In the *Corpus Hermeticum*, a series of philosophical dialogues that circulated together in late antiquity, the divine was given the name Poimandres, which itself has been speculated to be another name for Thoth. In addition to Greek and Egyptian influences, late antique Judaism also seems to have influenced the authors of Hermetic treatises, as is evident in the gnostic character of some of the philosophical works. It may well have been the contribution of Neoplatonic monotheists that made some of these Hermetic texts palatable to a number of patristic writers.[27]

The variety of Hermetic texts is quite large. Scholars of antique Hermeticism tend to categorize them, for the sake of utility, into philosophical (and religious) texts and technical works. Technical treatises included magic, astrological praxis, alchemy, and various medical and pharmacological texts. It should be underscored that this categorization is a modern product. It is unlikely that the authors of the Hermetic texts had, for instance, a clear distinction between magical texts and other technical works. While many texts, including those of Zosimos, were ostensibly based on dreams and visions, a wide variety of styles was employed by authors of the genre. Hermetic philosophy, especially the notion of cosmic sympathy, was the pole around which the technical Hermetica orbited.[28]

The *Corpus Hermeticum* puts forward perhaps the clearest statement of Hermetic philosophy and is placed at the center of the tradition, along with the Latin *Asclepius*, which exists only in fragments of its original form.[29] The *Corpus* is a series of dialogues, which, save for the first, are between Hermes and one or more of his students.[30] In spite of the Hermetica's reputation for obscurity and metaphor, the philosophy of the *Corpus* is quite coherent. The doctrine of cosmic sympathy emerges in the context of cosmology. The Hermetic universe is understood as having been created by the Divine, that is, Mind, which literally "thinks" the world into being via his *Logos*.[31] The expression of the world and the thought of the divine are transmitted to the sensory world by means of planets, sometimes called "emissaries." The harmonious revolution of the planets is indicative of the order brought by

Mind to the universe and, in a fashion recognizable to students of astrology, they infuse through their emanations various qualities and characteristics into all other things. The planets thus are linked to the idea of fate and destiny, common ideas to the Hellenic world, though the *Corpus* clearly implies that with *gnosis*, one can mitigate harmful celestial influences and activate that portion of Mind that exists within himself (and perhaps herself).[32] The planets themselves work through *daimons*, which are something like soulless middlemen who ensure planetary influence.[33]

Among the salient implications of this doctrine are two key ideas. First, divine power is inside the entire universe. The cosmos is alive, and all things are connected, even if such connection appears tenuous or does not appear at all. Second, the rotation of the spheres and other celestial bodies corresponds to a range of natural phenomena. The sun, for instance, is tied to certain plants, animals, and characteristics, as are the moon and other celestial bodies. Thus, one can utilize various quotidian substances—herbs, blood, stones—to leverage various qualities of the celestial agents. This is theurgy, the manipulation of matter to create wondrous effects. While theurgy is often characterized as a kind of magic, it is not opposed to the Christian world of late antiquity. As Peter Brown pointed out, the figure of Simeon Stylites, the column-perched hermit "bending the will of God by his prayers," is a fine contemporaneous example of the blurry boundary between theurgy and devotion in late antiquity.[34]

We can see evidence of cosmic sympathy in action in a later Hermetic text called the *Golden Compendium*.[35] This text is frequently tied to the *Kyranides*, not least because its putative author, Flaccus Africus, states that it too was found in the tomb of the legendary Trojan King Kyranos. The text is late—probably fourth or fifth century C.E.—and has been variously described as an herbal, an astro-pharmacological text, or a magical treatise. The range of categories ascribed to the text indicates that, properly speaking, it does not belong to any of these genres alone. Seeing the future, becoming invisible, making someone bark like a dog, and the like accompany more quotidian remedies like cures for bowel troubles, impotence, and skin ailments. It embodies the sympathetic doctrine, however, in justifying the efficacy of concoctions of remedies and potions. The way in which it was appropriated by some Western Christians is instructive for understanding the degree to which the elixir of Bacon and Rupescissa may have appeared structurally similar to the Hermetic concoctions, but still quite distinct from this tradition.

The *Golden Compendium* exists in both Greek texts and some copies based on a Latin translation made in 1096, placing it among the earliest Hermetic texts to be translated from Greek and Arabic during the long twelfth century. The 1096 text is also one of the earlier extant manuscripts in any language. I will follow this version in the discussion below, because the Latin text has more bearing on our discussion of the Franciscans' elixirs. The text opens with a discussion of its legendary provenance. It is addressed by Flaccus to his friend, the astrologer (*epilogistico*, lit. one who reckons), Harpocrates (who, like Hermes Trismegistos, was a god-turned-man).[36] Though the text predominantly focuses on the productions of medicines and potions, its significance to the astrologer is clear. The text describes seven plants associated with the seven planets, that is, various herbal concoctions that derive their potency from their relationship to specific celestial bodies.[37] These relationships, however, are not static, and the effects depend on planetary positions at various times. Therefore, herbs must be picked during periods when planetary influence is strongest. For example, the herb saxifrage, which is associated with Mars, must be picked in March. It is during specified times (a few days during the month), then, that the qualities of the herbs are at their most potent. The theory underlying the text is twofold. The cosmic aspect, if you will, is the efficacy of astral influences on the corporeal bodies. The sympathetic aspect pertains to the notion that like produces like.

This second notion is not always transparent, at least not to the modern reader, but there are some common-sense examples that demonstrate the relational aspect of the cures. For instance, a cure to ease ailments of the belly requires goat blood. Goats' ability to digest nearly everything was as renowned in late antiquity as it is today, apparently. Blindness is cured with herbs mixed with the blood of moles and nocturnal birds.[38] (Blood is a common and unsurprising ingredient. Hellenic culture long understood blood as the key to vitality. It flowed through the heart, which was frequently viewed as the home of the *psyche* or soul. As such, blood could be understood as conducting the vital essence of a particular creature throughout its form.) Other parts of animal bodies, such as stag horns, are employed only for ailments related to horns, such as bones or teeth.

To sum up, herbal ingredients are selected for their relationship to planets. Blood and other animal parts or secretions are selected based on the cure to be effected. Frankly, the question of whether these cures are "magical" or "natural" misses the point. These categories do not seem separable within the text itself. Clearly, there is a philosophic logic underpinning the text, yet the cures themselves were "revealed" and not based on empirical observation. It might be better simply to leave them as "Hermetic," with all the religious, magical, and proto-scientific characteristics that inhere in the term.

Did this Hermetic ideal, and the philosophical underpinnings of Hermeticism, survive the close of antiquity in the West? This is less clear. One survival may have been the *Mappae clavicula* ("The Little Key to the Chart"). If it was a Hermetic survival, however, it is clear that its medieval inheritors did not know this. The *Mappae clavicula* represents a tradition that first surfaced in Latin texts around the year 800 C.E.[39] Among the most famous of these texts is the *Compositiones ad tingenda musiva*, first catalogued by Muratori, which belongs to a series of works on the creation of pigments and dyeing. The twelfth-century examples of the *Mappae clavicula* have their roots in a parallel collection of ninth-century texts and exist as something of a "compilation of compilations" of the technical arts of the era.[40] The *Mappae clavicula* displays intricate knowledge of alchemical processes of its era, which is not saying very much, without any mention of alchemy itself. Moreover, it appears to be a compilation of recipes rather than a philosophical-epistemological treatise.

At first blush, the *Mappae clavicula* appears to lack any principle of organization or regard for duplication. Sections of the Latin are riddled with grammatical errors or oddities, a likely result of being literal translations from Greek. It includes various techniques and recipes of dubious value, and is far from an exhaustive catalogue of techniques available during the era of its compilation.[41] One manuscript contains eleven chapters before the prologue, and treats everything from making gold leaf, ink, soap, mortars, plaster, and a small incendiary, to cleaning silver and building a toy castle and soldiers.[42]

There remains, however, a contingent of scholars, including Bernhard Bischoff, Robert Halleux, and Paul Meyvaert, who have argued strongly for an understanding of the text as a vestige of late antique esotericism and alchemy.[43] Halleux and Meyvaert, however, have unearthed some clues as to the meaning of the text, which suggest an esoteric and alchemical heritage. They have noted that the terminology of the "key" has been mentioned in other alchemical texts, notably "The key to the art," attributed to Zosimos the Panopolitan by the tenth-century Arab indexer al-Nadim, and a "key of Hermes" mentioned in a letter by the Byzantine historian Michael Psellos (1017/18 and after 1078). They posit that, given that many of the recipes and techniques owe their origin to Greek sources, the title itself is likely the result of a literal translation of the Greek κλειδίον χειροκμήτων, literally the *little key of the cloth*, but more figuratively, the *little key of things made by hand*, that is, artificial things.[44]

Bernhard Bischoff, for his part, made an intriguing if inconclusive argument based on some textual surmises, which, if correct, would certainly merit the text's inclusion into the Hermetic corpus. Bischoff observes that the prologue to the text invokes Christianity, but conjectures that this is a veneer overlaying an older, pagan tradition. In the prologue to the work he argues that Latin "*haec mei/hec meis*" is a scribal substitution for *Hermaeis*, that is, Hermes Trismegistos, allowing Bischoff to connect the *Mappae* to the Hermetic tradition. While Bischoff is confident of his conjecture, he maintains it is indeed only a conjecture, though Meyvaert and Halleux follow his line of argument and expand upon it.[45]

Whether or not the *Mappae* was a garbled Hermetic survival, it is quite improbable that its readers would have recognized it as such. Whether or not it began life as a recipe collection, that is certainly what it was in the Middle Ages. Regardless where one stands on the evidence, the attempts to establish the text's esotericism are so laborious and speculative that it is clear that its medieval readers and copyists (especially in the twelfth century) were hardly steeped in Hermetic language, secrets, and codes. For the purposes of our discussion, the provenance of the *Mappae* is less important than the fact that it operated as a recipe collection, disconnected from any putative connection to the *Hermetica*.

The aforementioned *Golden Compendium* provides a similar example of how Hermetic texts were likely to be understood by their medieval readers, especially from the twelfth century onwards, when learned translators made unmuddled treatises available from Arabic and Greek. While we can pick out the Hermetic themes of the *Compendium*, they are not so obvious as to require any kind of major retraction or statement by a Christian author translator. Terms such as divine are used frequently, but without any specific religious connotation.[46] For instance, marrubium is "a heavenly and most excellent drug (*divino et excellentissimo farmaco*)," saxifrage "a celestial medicine (*caelesti medicamine*)," peony a "heavenly medicine (*divino medicamine*)," and sage "a heavenly plant (*divina herba*)."[47] The tone sometimes feels closer to salesmanship—treat your aches and pains with this heavenly herbal liniment!—than to any kind of theological claim. Heavenly is not simply synonymous with efficacious, though. The text does suggest that these medicines are special, but the efficacy of the herbs and the medicines in which they are components depends on more than the physical properties of the ingredients. One need not know of, nor adhere to, the doctrine of *sympatheia,* however, to understand the capabilities of these herbal mixtures. In fact, in a structural sense, the usage of the term divine is quite close to the way in which Rupescissa tends to use the term miraculous. Therefore, whatever the link of the *Golden Compendium* to pagan tradition, the verbiage within the text is adaptable to the concept of a Christian sacral universe.

Marginal notes on some of the manuscripts make this point more clearly. Lynn Thorndike noted that the text collection of the *Golden Compendium* he examined concluded with instructions that the gatherer of herbs reflect on the Lord's Passion while herbs are being collected.[48] This kind of devotional mindfulness appears in Bacon's work, as we will see shortly. Though the emendation of Christian mindfulness is (unsurprisingly) not found in all copies (and all marginalia are omitted in DeLatte's edition), it is in accord with the Christian sensibility of the text's readers. A marginal note from a different manuscript, for instance, tells its reader that the cruciform shape of the roots of the peony makes this herb particularly effective for casting out demons.[49] This is hardly the only mention of remedies against demons, but like the exhortation to be mindful of the Passion, it makes the link to Christianity explicit and recasts the religious matrix in which the text exists. The character of this text depends as much on the cultural context of its readers as it does on the material contained within it. One could say much the same about the *Secretum secretorum* used by Roger Bacon.

The upshot of this discussion is that while the Hermetic genealogy of these texts may have shaped them, this genealogy is often superseded by the context of the Western, Latin, medieval reader. Even when Bacon utilizes a text like the *Secret of Secrets,* he is doing so in a manner governed by the suppositions of a Franciscan living in the Christian cosmos described in Chapter 1. The difference can be summed up as the difference between a devotional approach and a theurgic one. The boundary between these ideas is both blurry and porous, but the comparison between them helps define what is at stake in a religious alchemical text.

Intention and Devotional Disposition

There is a subtle, but significant, difference in the original Hermetic attitude toward texts like the *Compendium* and the way in which medievals approached them. Another astro-botanical text, called the *Thessalos* after its putative author, helps make this distinction. The text begins with a letter to a king, a kind of foreword to the herbal remedies contained later.[50] In autobiographical fashion the letter recounts the intellectual (and physical) journey of the physician Thessalos. While the letter is a trope of Hermetic literature, it provides some interesting details as to the mindset of the Hermetic initiate. Intriguingly, much of the letter focuses on the great failure of Thessalos to utilize a corpus of Hermetic remedies ascribed to Nechepso (Lat. Notopsus; himself the putative recipient and author of other Hermetic texts). Branded a laughingstock for his failure, Thessalos wanders away from Alexandria, where he had been practicing. Refusing to go home out of shame, he wanders the ancient cities of Egypt, and finally happens upon an old priest grudgingly willing to aid him, after Thessalos tells him he is content to die if he does not have an opportunity to speak with a deity. The priest sends him off to fast while he makes preparations. Thessalos returns three days later, and the priest tells him he may speak to either a dead person or a god. Thessalos chooses the deity Asclepius, and sits down opposite the cathedra where the god will reveal himself. After a short time, Asclepius appears and comforts Thessalos that while he has called out to a god now, soon he will be called a god. Confronted with why Thessalos failed to carry out the instructions of Nechepso's botanical remedies, Asclepius tells Thessalos that while Nechepso was quite wise, he failed to account for the movement of the planets in his discussion of effective remedies. Asclepius thus perfects Thessalos's knowledge, which is passed on via the list of herbal remedies that follows.

The success of Thessalos might have something to do with devotion, but not in a Christian sense. Thessalos's success comes from forcing a meeting with Asclepius, not because he offers prayers of thanksgiving. Thessalos's "devotion" is also a means of compulsion. The narrative suggests that the exercise of Thessalos's will at the cathedra of the deity initiates the theophany. The demand to speak with a god is typical of theurgy, as is self-divinization.[51] The *Thessalos* serves two purposes, then. On the one hand, it provides practical knowledge of herbal medicines. On the other, it reinforces the religious aspirations inherent in Hermetic practice, by modeling both theurgic practice and the ascension of the initiate. Revealed knowledge proves transformative for Thessalos.

A very different devotional picture, from medieval Christendom, comes from a "medical" text tucked into a fourteenth-century Franciscan manuscript on the *ars dictaminis*.[52] The inclusion of various cures in nonalchemical or medical manuscripts was rather common. Friars, notaries, and scribes suffered the same ailments common to everyone else in the Middle Ages—styes, sores, coughs, fevers, toothaches, and so on—and often included recipes or advice for dealing with such ills in other texts as something of a nota bene to their readers. Keeping in mind the seriousness of even minor ailments in an era without antibiotics or much in the way of analgesics, one can see why such interpolations made their way into the marginalia or spare leaves of even the most valuable texts. In most cases, these consist of recipes or fragments, perhaps with an endorsement, but the vignette in Biblioteca Casanatense MS 9 stands out for both its narrative and its religious implications.

In revealing a "miraculous cure for the cure of ulcers, abscesses and even mortal wounds," the author of this medical vignette provides a colloquial account of how Christ himself revealed this knowledge to a trio of (ostensible) Franciscans.[53] These "three good brothers," on their way to gather herbs for their ailments, encounter Christ on their journey. He asks them their purpose and, upon learning that they seek cures for their wounds, Christ urges them on to the Mount of Olives.[54] The ingredients of their cure are simple, just some wool with which to apply olive oil to their wounds. The cure cannot be effected, however, without the three brothers also "speaking to and invoking God (*dicendo et coniurando*)."[55] The cures effected pertain to quite mundane, though serious, ills. (In addition to wounds, the text also treats worms, skin lesions, seizures, and so on.) The text presents an interesting dichotomy: miraculous cures in recipe form. Thanks to Christ's instruction, ailments will affect the friars as they did Christ, that is, not very much, as in the case of Christ's side wound: "Blood flowed from within, yet He did not suffer, the wound did not fester, nor suppurate, nor did it make a scab or a fever or a pustule."[56] The dichotomy also makes itself felt in that the cure given to the three good brothers is achieved through simple prayer (usually a *Pater noster*) combined with ritualistic elements such as genuflections.[57] Both the knowledge and the efficacy of the cures are due to Christ's goodwill, which the narrative suggests has been earned by the brothers through their selfless devotion to God. The brothers were grateful to God *before* they received their cure as well as afterwards. There is nothing to suggest that brothers' devotional activity was transformative in anything approaching the manner of the *Thessalos.* Rather, they were given encouragement from Christ and freedom from ailments that might impede them in their appointed tasks on His behalf in the mundane world.

The goal of the text, however, is not to laud the brothers, but rather to provide instruction for how to replicate the healing powers revealed to them by Christ. By following the example of the friars, mimicking both their actions and their words, one might benefit from these simple cures. In addition to the *Pater noster*, there is a prayer to be recited that is likely adapted from the Mass.[58] These are far from empty recitations. It is implied that the efficacy of the cures depends on the intent of the supplicants. The prayers and other aspects of the cure are to be carried out "*very devotedly* and *with faith* (*multum devote et cum fide*)" (emphasis mine). Intent matters here. It is not enough to follow precise instructions, since the cures occur based on God's will (*deo volente*) rather than that of the supplicants. Far from being secret knowledge, the text enjoins its readers "never to deny knowing how to do these things to anyone who wishes to know."[59] Therefore the cures are, like other miracles, aimed in all likelihood at bolstering the faithful with another example of Christ's beneficence and presence in the world. The tone of the text is devotional, however magical it may appear at first blush.

This devotional disposition was familiar to medieval scholars, who theorized about it in regard to the Eucharist, the central ritual of the Christian faith. Franciscan authors in particular highlighted the role of *intentionality*, namely, that spiritual benefits were conferred on the recipient of the Eucharist only if certain intellectual and motivational criteria were met.[60] Early Franciscan theorists, such as Alexander of Hales and William of Middleton, stressed the need for recipients to cognitively apprehend the sacrament. William did allow for the possibility, however, that unbelievers or unfaithful Christians who took the sacrament might in some way benefit from it. Unbelievers might be moved spiritually, and those whose faith is wanting could gain knowledge of God. The effects are more salutary for the faithful, however, whose charitable attitude toward the sacrament joins them with Christ in love. Writing after William, Bonaventure and later Franciscans such as Duns Scotus and Nicholas of Lyra continued to stress the importance of the recipient's attitude. Bonaventure highlighted two aspects of intentionality: the desire for the sacrament and the recipient's faith in the sacrament itself. There is little doubt that Franciscans were knowledgeable about these academic discussions, as most of our evidence comes from the Sentence commentaries of these Franciscan masters. Peter Lombard's *Sentences* was the principal theological textbook of the Middle Ages, and listening to masters or lectors comment on the *Sentences* was a staple of theological education. Thus, Franciscans argued that attitude and disposition played a central role in spiritual matters.

The importance of intentionality and the devotional mindset is underscored when Christian rituals are examined. The term "ritual" has no single, satisfactory definition (much like "religion," one of its parent categories—though rituals are often applied to very secular activities), though it needs to be pointed out straightaway that there is broad agreement with Catherine Bell's argument that "the features of formality, fixity, and representation are not intrinsic" to ritual.[61] These can be means of accomplishing ritual action, but do not define it. Hence magic can be ritual, but the repetition of a magical spell is not evidence of a person having acted ritualistically. The diffusive meaning of the term ritual has led some scholars to suggest doing away with it entirely, or focusing on ritualizing or performance.[62] Too often, there is little in the way of reflection on how this category is used, and it appears in a number of descriptions—especially descriptions of magic and liturgy, where ritual is simply defined as analogous to the practice—as though its meaning is unproblematic. I do not intend to offer any kind of magisterial definition here, however. As I intend to use the term, ritual is a specific kind of human activity that is inherently social in nature.[63] The fact that it is social does not mean it cannot be done while alone. For example, a medieval monk or nun might practice self-mutilation by him- or herself, but the practice has everything to do with the social matrix in which the monk or nun exists.

Rather than dealing with ritual as an abstract category, Franciscan liturgy provides us with a more direct means of approaching what the *habitus* of ritual life presented to its participants. Francis himself was an exponent of the importance of ritual, and he is credited with compiling divine offices for his own use and that of his brothers. One authentic liturgy that survives is his *Office of the Passion*, which was used five times during the year.[64] The Office itself is a compilation of various verses taken from a wide number of Psalms, with very few of Francis's own words inserted within the compilation. This is the same style as used by Francis to compile his early Rule, meaning that recourse to specific scriptural verses was careful and deliberate. Importantly, this ritual, unlike the Eucharist, carries no presumption of a miraculous result or particular supernatural effect, putting into clearer focus the importance of ritual.

In spite of the compiled style of Francis's liturgy, it has a progressive structure. The Office begins at Compline on Holy Thursday, iterating the tribulations suffered by the celebrant(s): "My wanderings you have not counted; my tears are stored in your flask; are they not recorded in your book?" (Ps 55: 9) Similarly themed verses follow, interspersed with plaintive prayers (also taken from the Psalms) for God to come to the aid of the celebrant. Picking up again at Matins, the liturgy rehearses the sufferings and alienation of the celebrants. At Prime, however, the liturgy takes a thematic turn: "My heart is steadfast, O God; my heart is steadfast; I will sing and chant praise." (Ps 56: 7)[65] At Terce and Sext, the themes of the first two hours repeat themselves. It appears that such will be the case at None, but then at the end of the liturgy, Francis returns to the triumphalism present in Prime: "But the Lord redeems the lives of his servants [with his own precious blood]; no one incurs guilt who takes refuge in Him." (Ps 33:23) The bracketed text is Francis's own addition to the Psalm, reinforcing the Christocentric interpretation of these Psalms in regard to the suffering (and deliverance) of the Passion. At Vespers, the triumphalism turns to celebration of the majesty and power of God. The slings and arrows lamented by the celebrants for most of the hours of the liturgy have been recompensed through the justice of the Lord.

The narrative progression provides a sense of what the liturgy accomplished. Despair from alienation and the hatred of fellow men leads to joy at the coming of God. The sufferings and wrongs are redeemed through Christ. This progression of the liturgy reinforces the ideal Christian concept of justice. But this is not a traditional narrative. It is a participatory one, where the celebrants begin in the real world—the world of slights, dangers, hatred, and suffering, and transition into what Seligman and his colleagues have called the "subjunctive world."[66] The subjunctive world, as the grammatical nomenclature suggests, is the world as it could or maybe *should* be. Rather than drawing a precise boundary between sacred and profane, the ritual world envisioned in this liturgy blurs the line, reframing suffering and displacement from individual experiences of privation to a collective experience. This is not an experience of the divine, but rather the experience of the tribulations of life framed collectively within a set of shared ideals. Here, in a liturgy for the Passion, participants are urged to identify their own sufferings with those of Christ during his Passion, and likewise to share in redemption. Seligman et al. posit that the ritual mode is central to the negotiation of the shared social world. It is necessary to reiterate here that the liturgy developed by Francis was designed to be participatory. Rather than a spectacle to be witnessed, the liturgy requires action and involvement. This is the medieval sense of *religio*, one's duty to worship—above all to *do* religion.

We can supplement this anthropological model by overlaying it with the model of ritual participation developed by medieval thinkers. I suggest it is useful to regard William of Middleton, Bonaventure, and other Franciscans as ritual theorists in their own right. In their eyes, intentionality was central to the success of the ritual, precisely because the ritual exists in tension with one's individual experience. Let us look at William's description of intentionality in the reception of the Eucharist in more detail, this time as a theory of ritual participation. William begins by saying that beasts receive nothing from the reception of the Eucharist. Beasts cannot understand the ritual, or even that they are participating in a ritual. Being outside the social matrix makes participation meaningless for them. Hence they eat only bread or drink only wine. Next come those who do not believe. These gain an aptitude or perhaps a propensity (*aptitudinem*) to see or comprehend the spiritual nature of their endeavor. Therefore, those who are outside the Church might not gain the full benefits of the ritual, but they may have an inkling that what they are doing is an important communal experience, capable of linking them socially to other participants through a shared experience. They are in a liminal position, but the door is open to fuller participation. Finally, when William of Middleton argues that only those who approach the sacrament in the fullness of faith receive union of love, he is not only speaking of the spiritual effects, but intimating the same idea that ritual theorists have—namely, that those who embrace the logic of the ritual and participate fully become enveloped in the subjunctive reality.[67] The union of love in Christ is shared among all the faithful, mediating and contextualizing individual experience through the shared subjunctive sense of their union with the divine. Whether or not modern scholars choose to agree that intention is important in ritualizing, medieval ritual theorists clearly did. Hence, when we find intention or devotional disposition in elixir texts, what is being suggested is that these belong to the same subjunctive reality as that found in medieval Christian ritual.

The Elixir as a Subjunctive Reality

The elixir belongs almost entirely to the subjunctive world. There is evidence that both Bacon and Rupescissa understood this to be the case, though that did not deter them from the belief that what *could be* could become what *is*. When approaching alchemy as a subjunctive, then, I am approaching it not in the sense of its belonging to the genre of magic, but in the way in which authors such as Bacon and especially Rupescissa envision the creation of the elixir as a manifestation of what God *could* do and a product that *could* exist within the framework of the world created by God. Here we wander into new territory. We know that the elixirs developed by Bacon and Rupescissa are not capable of the effects claimed by their authors. It is tempting simply to bracket this fact—as I have done for most of this book—and leave well enough alone with a thick description of medieval thought on the elixir. Instead, however, we must come to grips with the fact that neither Bacon nor John ever realized the elixir in the fullness of his hopes, though both men claim to have seen the elixir used. Can we simply pass this off as gullibility? The evidence suggests we cannot.

Glossing the text of the *Secret of Secrets* prior to fully developing his own alchemy, Bacon tells us that he has seen "a most experienced physician (*peritissimum medicum*)" use the inestimable glory. The person cured was likely Alphonse of Poitiers, Prince of France and younger brother to the king.[68] The cure not only resulted in the restoration of physical health, but also rectified his mind and character, purging avarice, melancholy, sadness, and "many other ailments of body and soul." Thereafter the prince became "extraordinarily generous, bold, and high-spirited," having been cured of "moral failings (*viciis . . . moralibus*)."[69] Bacon goes on to tell his readers that while hardly any physicians use the *Secret of Secrets*, the inestimable glory can be made by extremely learned apothecaries. The inestimable glory, as it is treated here, is a variant of the theriac, with Bacon even going so far as to reference Montpellier as a locale for finding the proper ingredients. Bacon does not state how he comes by this report, however. Bacon was in Paris during at least part of the time of Alphonse's illness, but it is not clear how he knows, for instance, that the physician treating the royal family utilized the (very vague) instructions of *Secret of Secrets* to make inestimable glory, nor that he added to it herbs and various animal parts.[70] It is possible that Bacon managed to interview the physician, but it is also likely that he is surmising that the great recovery made by Alphonse had to have occurred because of the use of a spectacular medicine. Bacon's description of the inestimable glory comes across as somewhat mundane in this passage—it is not as fully developed in his gloss of the *Secret of Secrets* as it will be in the *Book of Six Sciences*—but I would argue that the naturalistic qualities of the elixir do not detract from the fact that Bacon understood his elixir as a subjunctive reality.

Naturalism does not preclude subjunctivity—quite the contrary. The world that could, or should, be is the world governed by both divine and natural law. In the case of the medieval Franciscans, natural and divine laws flow from the same place. Recalling our discussion of Bonaventure in Chapter 1, Bonaventure understood the natural world as the first divine book—the way to understanding God before the fall robbed human beings of their ability to see the world unclouded by their own sinful and imperfect natures. God's justice and love were inherent in the created world—a world that is apprehended through physics, optics, and other sciences. Bacon, in his own way, agreed. At the very least, Bacon believed that scientific knowledge of the world was necessary for a full exegesis of scripture, and more generally he articulated scientific disciplines as the first steps on the ladder of knowledge to understanding God.[71] The subjunctive world—the world of possibility—was the world where divine and natural laws worked the way in which medieval observers thought they should. Given the disjunction between medieval understanding of natural laws and what nature does, the subjunctive world might have been a more real world than medieval people would have experienced. Indeed, Bonaventure hammers home the point that human sin has occluded the human ability to see the world for what it really is.[72] There is certainly a Platonic element to this kind of thought, but Bonaventure and fellow Churchmen were much more confident about what that underlying reality was, because although they may not be able to read the Book of the World, sacred scripture was handed down in order to repair human understanding. Rupescissa, for his part, seems quite sanguine regarding the apprehension of the subjunctive world. He states that the world is so awash in *mirabilia* that one can hardly think or speak of them without babbling.[73] The subjunctivity of these wonders or miracles is alluded to by the fact that they are not only concealed, but also capable of being revealed. We cannot know the full list of miracles God has placed across the world, but they are always present.

The point to underscore here is that subjunctivity does not equate to unreality. Rather, in the sense meant by Bonaventure, Bacon, and Rupescissa, it suggests the potential of reality to be in harmony with what scripture or theological truth claims suggest. Thus, for someone like Bacon, the truths gleaned from Aristotelian observation and classification had to exist in concert with both scripture and the changes to the natural world suggested by the presence of an omnipotent deity active in the world. Bacon's Aristotle was not just the Aristotle of the *libri naturales* or the *Nicomachean Ethics*. He was also the Aristotle of the *Secretum secretorum*, who explicitly stated that God intervenes in the world as a result of human devotion (prayer, fasting, and so on).[74] Bacon amplifies this idea in his own work when he speaks about the prayers of women, of whom he has a rather low opinion even for the standards of his day. "Even a poor little old woman with her prayers and merits, if the kindness of God is at her side, is able to change the order of nature."[75] Bacon is suggesting here that Christian devotional activity can achieve miracles. This is hardly a novel idea, but Bacon is clear that God's activity in the world frequently changes nature or goes against natural law altogether. God can change a person's complexion or can make a storm simply disappear. Bacon's subjunctive world is where divine justice and divine mercy trump the realities of day-to-day existence—the same theme found in the Franciscan liturgy discussed earlier. What remains is whether Bacon understands the elixir itself to fall into the same category.

The evidence is suggestive, but not conclusive. As described in Chapter 2, Bacon's elixir is a naturalist concoction, made from the elements. What is more, Bacon justifies the utility of the elixir by the fact that a variety of natural substances are found that have similar effects. At the same time, it is not clear that natural occurrences per se fall outside his understanding of the miraculous. In the *Tractatus brevis*, Bacon's introduction to the *Secretum secretorum*, he holds to the standard definition of miracles as events or effects that are above or contrary to nature. He even debunks many alleged miracles. A good example of this is his description of the magnet, discussed in Chapter 2, but he also explains away widely attested miracles. For instance, he argues that a Norwich woman's abstention from food for twenty years, far from being a miracle, was due to the influence of a constellation that had equalized her complexion.[76] Obviously, this is less a product of some kind of proto-skeptic mentality than a sign of Bacon's agenda to put forward the potency of *scientia experimentalis*.

Given Bacon's desire to enfold nearly every putatively magical or miraculous activity into *scientia experimentalis*, Bacon's "naturalist" explanations are not as straightforward as they seem. In the course of debunking a wide array of incantations and magical inscriptions, Bacon admits they may well be magical in nature, or deceitful, but are not in and of themselves necessarily opposed to Christian doctrine. In his *Letter on the Secret Works of Art and Nature*, for example, he takes care to note that certain ritual activities and prayers, generally ordeals, are licit when performed under the auspices of clerical authority.[77] Bacon also argues that "the astronomer is able to form words for chosen times which will have inexpressible power," namely the power to almost completely control natural forces as well as people.[78]

Uttering or inscribing words of power is a *scientia* of incantation and magical inscription.

A closer look at Bacon's understanding of prayer further complicates the issue. In the *Tractatus brevis*, Bacon offers something like a history of prayer, which emerged not in some primordial biblical time, but from pagan philosophers in direct response to (negative) astrological foreknowledge. Bacon reasons that since the heavens have been arranged by God, certain non-Christian philosophers who had a measure of God's grace (such as Plato, Aristotle, and Avicenna) "established prayers and sacrifices concerning worship of the divine."[79] Bacon argues that these prayers and sacrifices were not demonic nor inherently opposed to Christianity and in fact resembled those devotional activities of the patriarchs before the time of Moses. While this is hardly a satisfactory historical argument, it reinforces Bacon's claim that prayer is meant to be an active force in the universe and is consistent with philosophical principles. If we look at the part of the *Secretum secretorum* that Bacon is glossing here, we see that pseudo-Aristotle recommends fervent prayer and devotional activities whenever the stars suggest that bad times lie ahead. If calamities cannot be averted outright, they can at least be mitigated.[80] Bacon, however, does not appear to be suggesting that prayers are an ex post facto activity for astronomers. Rather, Bacon suggests that prayers are part of the astrological investigation: "Christian astronomers ought to use pious devotion and prayers to God and the saints and to invoke their aid in all their works."[81]

At first blush, there seems something akin here to contemporary athletes making the sign of the cross or pointing heavenward before or after a critical moment. Bacon has a more involved and liturgical kind of prayer in mind. Bacon advocates that Christian astronomers repeat biblical prayers as well as make inscriptions (*karacteres*), such as the sign of the cross, as well as figures of the Virgin and of St. Dionysius, whom Bacon regards as something like the patron saint of astronomers.[82] While astrology is not a ritual, Bacon is suggesting that the best results are achieved when the astrologer behaves in the ritual mode, focused on reproducing the right intention. The result of this devotional activity is that an astronomer "ought and can produce certain harmonious works in such a way that what he intends happens more easily and better."[83] This suggests that one might achieve astrological knowledge without such prayers, but Bacon nevertheless instructs astronomers to take a devotional approach to their work (*operans habet devocionem*). Here we see an iteration of the devotional disposition discussed above. Right astrological practice requires a certain kind of intent on behalf of the astrologer.

Astrology as a devotional practice takes on special meaning when we examine it in the context in which Bacon raises it, namely discussing those non-Christian astrologers who had "honored God according to the grace given to them."[84] It seems quite clear that Bacon regards their grace as manifest in a specific gift—namely their wisdom. The intention of devotion is not incidental to the practice of astrology, because God's grace is always part of the astrological equation. Just because God's presence is not made manifest through an act contrary to nature, that does not mean it is absent. This fits fluidly, moreover, with Joachim of Fiore's similar understanding of the grace he and others received—the grace of intellection, the ability to discern and understand. The dichotomy of natural/supernatural, mundane/miraculous misses the point of the creation of Bacon's elixir. It is no less "religious" simply because it is natural. Rather, it is natural in the sense conveyed in Chapter 1. It is consistent with a world created by God that does not require escape or avoidance, but one whose fundamental goodness was obscured by human blindness. Therefore, Bacon's elixir is a subjunctive reality because it *could* be realized in just such a world.

The elixir has been theorized, to be sure, as a naturalist compound, and various effects of the elixir are present in piecemeal fashion, but the awesome power of a Christian-making elixir has yet to be released. Bacon's ultimate secret is in many ways still a secret to him. There is a natural means of extending life—found in the case of a fortunate farmer who drank a golden liquid, a learned German man with papal letters attesting to his great age, and biblical precedents. There are the astrologically affected people—everyone in an individual sense, but also peoples whose complexions (as a group) must be explained not only by the regional variations of their homeland, but their different position in relationship to the heavens. Therefore it make sense to Bacon that the elixir can make these changes on a grand scale, but clearly he is not claiming that someone has already done so, else why write the pope a series of three treatises calling for the alteration of the educational curricula toward the *scientiae*? God's plan for victory over Antichrist, however, *should* allow for the creation of a naturalistic compound fused with celestial power that will reinforce Christendom in its most desperate hour. It could not be made, however, without the grace of God conferring wisdom and foresight among the righteous.

Like Bacon with his elixir, Rupescissa implies that the quintessence can be replicated by anyone, since the quintessence is derived "with human artifice."[85] One finds in his text many of the entreaties and injunctions aimed at secrecy found throughout the alchemical tradition that also suggest that one may simply carry out the process(es) he lays out and arrive at the same result. Likewise he voices concern that contemporary philosophers and physicians, whose only concern is profit, might gain knowledge that should be otherwise restricted only to the worthy.[86] This complaint is also common to other medical-alchemical texts of the era, in which the alchemists perceive themselves as possessing greater wisdom than their physician counterparts.[87] Thus, for all John's assertions that he is making miracles, these passages suggest that the wonders of alchemy are not in any way restricted to the faithful or even to Christians. Yet, the method by which John came by the knowledge of the quintessence complicates this picture.

John's alchemical knowledge is revealed knowledge. Here we find a similarity with his prophetic works, especially the *Liber secretorum eventuum* (*Book of Hidden Events*).[88] In the *Book of Hidden Events*, John bases his ability to discern the future on divinely inspired reading of scripture rather than on direct revelation. John makes it clear in his prophetic works that his rapture provided him with the means to understand (*intellegere*) the fullness of scripture, including the coming of Antichrist.[89] His prophetic ideas emerge from a mixture of study supplemented by divine gifts, and he describes his alchemical knowledge in similar terms. He tells us first that he studied philosophy for years during his ten-year term at the University of Toulouse, but this wayward period granted him no major insights. Rather, like Joachim, he credits God with having "revealed to me many secrets of natural philosophy."[90] This is the Joachite idea of the grace of intellection.

Divine revelation through intellection provides John a surety that his knowledge is correct. While any vision from God would contain true knowledge, one must still cope with the thorny problem of whether or not the visitation was true, a question often asked of female visionaries and certainly of John. Gifted with knowledge and understanding, John not only can assert truth, but can prove it. In the case of his prophetic works, this is done via Joachite exegesis. In his alchemical works, it is done through demonstration, laying out alchemical principles and processes. John provides all the knowledge and tools necessary to create the quintessence alchemically. John knows he will not be believed, so rather than just asserting a recipe for the elixir, he illustrates the theory of its potency, the various methods by which it might be made, and the uses to which it can be put.

John's concern for secrecy is a discordant note in what is otherwise a very clear and open account of the elixir. While John's role is magisterial, he treats his audience not as apprentices, but as fellow travelers. The evangelical men for whom he writes are just as worthy of knowledge of the quintessence as John is. He strives to persuade as much as instruct. For the most part, John's alchemical works are quite open and do not rely on ciphers or codes to obfuscate their contents. He speaks "without parables (*sine parabolis*)" and even goes so far as to decode texts of other alchemists.[91] Everything from ingredients to the design of an alembic to the distillation method itself is described in detail.[92] For all John's putative concern about secrecy, there is nothing about *De consideratione* that suggests he is deeply concerned with the text's diffusion—unless it is a concern that it not circulate at all. John's concern over misuse is more likely a device meant to entice readers with the promise of a powerful secret. His reliance on divine providence and on the prudence of his intended audience to preserve the quintessence comes across as more authentic than his fear of misappropriation.[93] This is supported by a claim that John makes at the end of the first book of *De consideratione*, where he tells us that knowledge of the quintessence is not the limit of alchemical powers. One could learn, for instance, all the hidden things *(interiora)* of nature as well as the ability to transform it. This is granted, however, only to a few, namely, those who have lived the holiest and most contemplative life.[94] Thus, as powerful as the secret of the quintessence may be, greater knowledge is available to those whose devotional dispositions are unmatched.

Another clue to reconciling the seeming contradiction between the divine origins and mundane production of the quintessence can be found in Arnold of Villanova's commentary on the pseudo-Joachite tract *De semine scripturarum.* The thirteenth-century Catalan was not only a prominent physician, but also an ardent Joachite expositor. Like John, Arnold did not conceive of a division in his thought between his apocalyptic and medical works. Science, as Arnold conceived of it, was very much a divine act. Rather than seeing the influence of God in healing alone, Arnold idealized the unlocking of the secrets of the universe as a profoundly religious activity that could "reveal prophetic knowledge."[95] The *Tractatus parabolicus*, a pseudo-Arnoldian treatise John knew, reflects Arnold's views in a discussion of alchemy. It proposes parallels between the Christian faith and alchemy, positing itself as "a biblical exegesis of the alchemical system."[96] Here, Christ is a symbol—through his resurrection—of alchemical transmutation. Thus, the *Tractatus* provides a theoretical model for alchemical practices that is not just based on the Christian universe, but is one in which an understanding of Christian truth is necessary to grasping alchemy.[97]

For Arnold and John, this theory was an expansion of the Joachite exegetical epistemology. Not only could one mine the scriptures for hidden knowledge, but the natural world itself became a text to be read, studied, and interpreted. Along with the apocalyptic resonances of John's alchemical work, this application of Joachite exegetical principles serves as one of the foundational links that bind together John's prophetic and alchemical works into a unified corpus. John's elixir shares with Bacon's, then, a sense that while its ingredients are found in the mundane world, God is involved in the use to which the elixir is put. As God's means of producing miracles in the world, the quintessence performs "just as [God] desires and not beyond."[98] John's equivocation on whether or not the elixir requires God's assent is a further sign of its subjunctivity. It is not supernatural, but it is part of the world as it could or should be. It *should* exist just for evangelical men. It *should* work in accord with God's will and not beyond. Most important, the quintessence, the essence of heaven, *should* be found in every earthly substance, a sure sign of God's goodwill to his creation.

John recognizes that the world as it is, rather than as it should be, does not offer the quintessence. "But you will say that all things which are corporeal in this world, and are for the use of the body, are elements or are compounds of elements. Therefore the root of life . . . is not able to be found by men in this world or in this age."[99] Yet, thanks to the beneficence of God, the quintessence does exist. It is the "*human heaven* that the Most High created for the preservation of the four qualities of the human body, just as he created heaven for the preservation of the entire universe" (emphasis mine). There is nothing that expresses the notion of subjunctivity more than the union of heaven and earth. A body adorned as a human heaven, rectified and purified by God, is as clear of a statement of the world as it could be as one is likely to find.

The Subjunctive Soul

Throughout this book, I have sought to convey how deeply held religious convictions (particularly regarding the end times) drove alchemical pursuits. While I think the emotional power of religion is largely (and perhaps too readily) taken at face value, it is perhaps also too easy to conclude that the advance of a scientific discourse is the product simply of rational argumentation (however much historians of science have demonstrated the contrary). The search for the elixir by these Franciscans was colored by personal convictions about religion *and* science, nowhere more so than in the case of John of Rupescissa, whose own sanity and salvation hung in the balance. For Rupescissa, subjunctivity is not only a means of theorizing the elixir, it is a strategy of survival.

Among the pejoratives attached to John's name, idiot (*ydiota*) must be among the least deserved.[100] *De consideratione* shows a keen and self-aware mind at work. Throughout *De consideratione*, John uses the term *phantasticus* (delusional, mad) thirteen times. *Phantasticus* was, rather than *hereticus*, the finding of the Cardinal Inquisitor who sat in judgment of John at his final trial in Avignon. It is not a term that I have found in the *Book of Hidden Events*, which was written for the trial itself prior to John's sentence, and his adoption of the term must be intentional. The places within *De consideratione* where the term surfaces suggest both John's initial discomfort with his sentence and his attempt to use the term as a shield for his continued writing.

In the *Book of Hidden Events*, a confident but respectful John tells his chief inquisitor to "judge whether it was almighty God himself who opened understanding for me . . . or whether I was deceived by an unclean spirit."[101] By the time he completed *De consideratione*, years of imprisonment had made John bitter and suspicious, and perhaps more politically astute as well, since the term that mocked and constrained him also freed him to pursue what for sane men was foolish or impossible.

In Chapter 2, I noted John's description of the limits of the quintessence, namely, that it could not confer immortality. One of the reasons for the awkwardness of the argument of a certain human shelf life was John's desire to confine his search to the realm of orthodoxy. There is another possibility, however, which is that John was merely paying a thinly veiled lip service to this idea. At the end of the first book of *De consideratione*, John expresses his inability to say more about transmutation because he is compelled by obedience and canon law. While the anger rings true in this passage, the claim about not being able to discuss transmutation is rather odd, especially since John authored, probably in 1350, the *Liber lucis*, a treatise on transmutation.

Returning to John's argument that the quintessence cannot confer immortality, he offers a clue to what may have been a streak of defiance cloaked under the rhetoric of obedience. The passage is so significant that it merits quoting at length. Immediately after citing Paul's letter to the Hebrews 9:27 ("It is appointed for men to die"), he says,

> Therefore it is mad (*phantasticum*) to work to find in this mortal life something able to render our bodies immortal from a mortal state. . . . Therefore it is mad (*phantasticum*) to say that God would give something to Adam outside Paradise—from which he cast him out, lest taking from the fruit of the tree of life he might live forever—through which he is able to live forever. . . . Therefore such an aid for prolonging life would be mad (*phantasticum*) and foolish.[102]

Structuring the passage rhetorically, John sets up a scriptural or theological objection against either the existence of the quintessence, or at least the attempt to obtain it. In every case, second-guessing either scripture or orthodox theology is an activity reserved for the mad. The significance of the term *phantasticus* cannot be overlooked, since John could have chosen a variety of more common Latin terms had he intended the passage to be anything other than self-referential. The explanation for the multiple occurrences of the term *phantasticus* may be as simple as John avowing his own sanity, but they may also be a tongue-in-cheek evasion over new questions regarding his doctrinal orthodoxy. By reminding his readers of how crazy it would be to find an elixir of immortality, he also reminds them of his own sentence. This could well afford him the assurance of being thought a liar or raving madman by his enemies, which was at least better than the stake. More subtly, it could be John's means of hinting to his true audience, the evangelical men and his other supporters, that he is providing them with knowledge that the corrupt potentates of Avignon would rather not release.

Something more than a mere declaration of sanity or truthfulness seems to be at stake when we examine exactly what it was that John says it is mad to do. Following closely on the heels of the passages cited previously, John cites yet another "mad" idea, namely that it would be crazy to think one could concoct a cure-all or elixir from anything that is corruptible—something that can rot, erode, break down, be transmuted, and so forth.[103] If any action could change the character of a substance, then it would technically be corruptible. Hence, from a purely logical standpoint, John must latch onto the incorruptible substance, namely the heavenly quintessence. This distinction, however, is at best a technical one, since like William of Ockham, John conceived of the quintessence as malleable.[104] For John holds that we can find the quintessence in any earthly thing, all of which are corruptible. We also know it ignites and can be consumed in that manner, something that should not be possible if the quintessence is not subject to human activity. Moreover, the quintessence amplifies various effects when combined with particular compounds, implying that some sort of mixture is made. Despite John's claim to the contrary, the quintessence seems very like a changeable substance made (at least in part) from corruptible elements. John's terminology suggests that he is aware of some of the contradictions present in his argument—ones that might lead various parties to scoff. But John had his supporters as well, whom these passages seem to address. Those who looked past his sentence and his captivity and believed in his prophetic utterances should also pay heed here. For many of the *fantastic* ideas appear to be exactly what John is pursuing. Later in the text, he tells his readers that if they seek the quintessence, they too will be "reputed to be mad (*phantastici*)," especially by philosophers ignorant of this new divine knowledge.[105] Certainly this can be construed as a warning that his readers may share his fate. After all, John often expressed anger at his ignominious treatment, but the constant invocation of madness at the opening of the first book may also be an invitation to join John in doing what is said to be impossible.

There is more evidence that this is in fact John's strategy. First, John draws a distinction between what he can say without penalty, and what he really knows. As noted in the fifth chapter of *De consideratione*, John bitterly states that he must hold back additional knowledge for fear of persecution. Later, in the second book of his treatise, in a chapter on how to drive off demons, John says that his remedy is "orthodox (*catholicum*)," but that "every remedy that is prohibited by the Church must be avoided."[106] The implication, of course, is that other effective remedies exist. Also, John occasionally allows for the possibility that one can *do* the impossible, despite how *fantastic* his claims may appear. This occurs notably in a chapter on plagues.

Citing Deuteronomy 28, John allows for the fact that God, in His righteous anger, sends plagues that cannot be cured.[107] And, of course, "if you presume to cure them, it would be attacked as both *fantastic* and insane."[108] After this, however, John provides a fairly lengthy list of ailments and their various astrological influences. This list includes plagues and illnesses resulting from the influence of Saturn (leprosy, catarrhs, consumption, and so on.), Mars (fevers, St. Anthony's fire, anthrax, pustules, and so on), Mercury (buboes), and the Moon (epilepsy and nightmares).[109] What is important about this list is that in it John connects the etiology of a host of plagues to the influence of the planets. The quintessence, however, nullifies these effects and wards off "universal peril."[110] While John never claims to be able to cure a person directly afflicted by God, since the quintessence mitigates the negative celestial influences that would cause disease the implication is that it is an effective remedy unless God is directly causing the affliction instead, without making His will felt via celestial influences.

At the same time, there is some evidence that John's relationship to the term *fantastic* was not simply as a rhetorical cudgel to shield his work. In the slim collection of specific cures that make up John's second book of *De consideratione,* twenty illnesses have been singled out as worthy of specific discussion. Of these, four cures deal directly with madness and its causes, and two deal directly with demonically inspired madness, including "fantastic passions, fancies, follies, infestations and temptations of demons."[111] This is the precise ailment John had dared the Cardinal-Inquisitor to ascribe to him in the *Book of Hidden Events.* Lest we think that demonic infestation is somehow ancillary to the other curative attributes of the quintessence, John reveals at the outset of his entire work that his atonement for past sins includes the following objectives, "to heal with divine goodness, to wondrously ward off impediments to holy prayer and meditation, and even to resist temptations of demons."[112]

John is familiar with contemporaneous scholastic opinions on demons, especially the claim that they are noncorporeal beings.[113] John realizes that this creates a problem for him, in that he wants to use a corporeal substance to affect an incorporeal being. John turns to scripture to make his argument. In the sixth chapter of Tobit, the angel Raphael, after saving Tobit's son, Tobias, from a giant fish, bids the young man to remove some of its organs. The angel proceeds to tell Tobias how they can be used to treat illnesses, and, of course, to drive away demons. John uses this episode to validate his claim that God has created in material objects not only the power to heal, but also to drive out evil spirits, precisely the function of the quintessence.[114] John does not need, then, to contradict the claim of incorporeality of demons. By citing scripture directly as well as Augustine's writings on the soul rather more obliquely, he sets up a credible, if thin, defense against scholastic arguments claiming corporeal measures could not be taken against demons.[115] Such "weak conjectures," he tells us, are not to be believed.[116]

Ultimately, this scholastic exercise serves as something of a precursor or even a straw man, since John's premise rests not so much on the various scholastic and scriptural surmises he makes, but on "what manifest experience teaches" about demons.[117] This is not wholly surprising, since the alchemical tradition, however much its adherents strained to link their work to extant knowledge, relied on some manner of epistemological empiricism.[118] John, of course, never says it is his particular experience that he relies on, though he has witnessed (or effected) cures of demonic activity.[119] John's discussion of demonic activity relies on the effect of demons on the humors, especially those having to do with melancholy and "unbelievable sorrow (*tristitiam phantasticam*)." This is part of a pattern of tying the etiology of madness to demonic influence by rehearsing the commonly held belief that the incursion of a spirit often preceded or accompanied "natural" changes in physiology and character. For instance, melancholy is ascribed to the buildup of damp and cold humors, since an abundance of black humors makes the bowels a delightful place for demons to dwell.[120] The bowels also make a fine base from which demons can further disturb the mind. As an interesting note, John demonstrates little inclination to gender madness, especially melancholy, which was understood to be an ailment predominantly of women. In any case, John's cures for madness and for demonic influence bleed together: the quintessence expels black choler and black bile (humors related to madness) and "takes away disturbances of the mind, lightens the sad heart, cleanses the brain and all one's faculties, leads to happy thoughts, removes the trick of demonic temptation, and destroys hopeless imaginings. Also it makes one forget every evil and restores men to natural mind."[121]

John's association of melancholy and sorrow with the term *phantasticus* is related to, but distinct from, the way the term was likely used to describe him, and somewhat different from the way John usually uses it. The term was hardly widespread, and could be taken as a synonym for mad—something of a catch-all included with words like *hebes*, *sensu carens*, *ydiota*, and French terms such as *lunatic*—or more nefariously in German as *truglicher* (deceitful).[122] The papal curia's use of the term in reference to Rupescissa seems closer to the classical definition and the root term *phantasia*. To use a contemporary turn of phrase, the papal curia essentially said John was "seeing things." John himself invokes this meaning when he discusses "fantastic apparitions" caused by a fever. John instructs his readers to weigh carefully (*perpendes*) the cause of the illness by evaluating the type of visions. If they are black or dark, then they are caused by melancholy; if they are golden, then they are caused by a fever and excess of the sanguinary humor.[123] In either case, John imputes no other value to the visions than a diagnostic one. It is worth pointing out, however, that John himself describes one of his own visions in a manner that might lead one to believe it was melancholic in origin. John relates that he had "a horrible vision in the night" wherein "a black beast emerged from the east." A parallel passage has a different description, all of which points to the subjective nature of evaluating one's own visions or the visions of others.[124]

While melancholy or black moods might fall under the broader definitions given to *phantasticus* as a sort of madness, this is far from the primary meaning. Rather, John seems to use *phantasticus* in connection with melancholy and sorrow as a means of separating normal sorrow from a sorrow caused by illness. The question is one of degree, not whether one is real or not. This same meaning occurs when John discusses "fantastic plagues."[125] Certainly John is not implying that either melancholy or plagues were merely imagined. Rather, they are "unbelievable" in their severity. This distinction implies that John was trying to domesticate the term. Perhaps co-opting the term was achievable, given the fact that *phantasticus* was neither a common legal word nor a "diagnostic" one. We have already discussed John's reliance on the term *phantasticus* to describe a number of ideas, tasks, and remedies in *De consideratione*. In these cases, as with "unbelievable sorrow" and "unbelievable plagues," the remedies John produces might not be believable, but that hardly detracts from their reality or effectiveness. The problem of being "unbelievable" is not a fault of the *fantastic*, but rather it is the failure of others to comprehend the truth.

One problem John creates for himself is that many of the *fantastic* ailments that the quintessence remedies generate false visions. As I just noted, the character of the visions can be a clue to the humoral etiology of the illness. But what else do we know about John's own visionary experiences? Are there clues in his visionary treatises that might inform us as to whether John suspected himself to be deceived? Fortunately, John does provide us with some evidence of how he defined visionary experience in general as well as the nature of his visions. First are two accounts of his own visions. In the *Book of Hidden Events* he recounts quite dramatically a vision following the preparation of his body through intense deprivation. The passage is worth quoting at length:

> To the most reverend father and lord in Christ, Lord Guillaume, Cardinal of the most holy Roman Church . . . I, John of Rupescissa, of the Franciscan Order . . . by your command have described the sequence of noteworthy future events revealed to me in prison as

> well and truthfully as I can remember. The mode of revelation was this: when in the year of our Lord Jesus Christ 1345, I wept for many days bound in irons in a prison of filth in the convent of Figeac, overwhelmed and wondering why I should have been sent with such cruelty by Brother Guillaume Farinier, then minister of Aquitaine, into prison which was full of dung and filth. Why should God have allowed me to be handed over into the hands of such cruel masters at the testimony of lying and perjuring men (as the Day of Judgment will show them to be)? It was then given to me to understand clearly that I had fallen into such temptation since I was about to reveal to the world Antichrist, and his people and his country, before he appeared as Antichrist. And I had been confined in the strongest afflictions and temptations, which I already suffered for five years, in order that I should be made suitable to understand the future. By this understanding I strengthened my prayer, made my penitence the weightier, and increased my vigils.
>
> And it happened that in the same year in the month of July, near the feast of Saint James the Apostle, when I was awake (*vigilans*) and praying, standing on my feet in the middle of my cell, holding a rod in my hand, immediately in an instant—*in the blink of an eye* (1 Cor 15:52), my understanding was opened and I understood the future with an intellectual vision which consequently will be described summarily and in brief.[126]

There are some notable elements in this passage that require emphasizing. First, John's divine understanding is *intellectual.* Second, he is awake (*vigilans*) when the vision comes, underscoring that he is even standing up, lest he mistake a dream for a waking vision. Robert Lerner has pointed out that John's adherence to the spirit of understanding rather than the spirit of prophecy was a critical theological distinction, owing to the fact that the spirit of prophecy was no longer a divine gift after the completion of the New Testament (though John would retract or nuance this claim in later work).[127] Third, John actually has two separate moments of understanding in this episode. The first understanding is that he is suffering in order to be prepared, with the effect that he voluntarily increases his vigils and resorts to further bodily castigation (the rod). In the *Liber ostensor* he tells us that he suffered "like a martyr of old."[128] Only after further suffering does John receive his second understanding regarding the coming of Antichrist.

John relates two other visionary experiences in the *Book of Hidden Events*, though neither is as dramatic. He had a prior vision of future events in 1340 on the day of the nativity of the Virgin, when he and the choir had begun the *Te Deum.* Again his understanding was opened "in an instant."[129] He also received a vision in 1346, then in a Toulouse prison.[130] This vision is notable because it adds an element to John's prophetic experience. In addition to being granted foreknowledge, he tells us "the weight of heavenly sweetness and of the presence of the great glory of God overwhelmed me."[131] Though John stops short of the beatific vision, he is being granted knowledge a well as an experience more akin to mystical ecstasy.[132] Finally, John speaks in the *Liber ostensor* of a 1349 true vision (*visus*). In this case John dispenses with the language of intellection, saying instead, "An angel of the Lord appeared to me in a vision."[133] The angel tells John that soon he will be freed from his current prison, and John is delivered in short order so that he can make his way to Avignon to plead his case. Indeed, in the *Liber ostensor*, he describes his Toulouse vision (as well as numerous others) with vivid imagery absent from his earlier descriptions.[134]

In spite of all his foretellings and divine visitations, John rejected the title of prophet.[135] Fortunately, he offers us some clues to interpreting this seeming contradiction. In the eleventh treatise of the *Liber ostensor*, John spends a great deal of time considering the fate of the various religious orders, especially his own. Earlier I referred to a number of places in which John critiques the Spiritual movement and seeks to separate himself from other putatively more radical heirs of Olivi. In addition to offering descriptions of the persecutors of the evangelical men within and without the Order, John offers some details on the behavior of the elect. There are ten grades of sanctity, many of which have their own subdivisions, and it soon becomes clear that the elect and John have a great deal in common.

Some clues to how John understood his own visionary experience come during his discussion of the fifth grade of sanctity of the elect, *illumination* (*illuminatio*). Not sparing with symbolic language, Johns says the quality of illumination has seven *rays* (*radii*).[136] The first ray is the demonstration of the perfection of the life of Christ, and the second is the conversion of the world through said demonstration.[137] The third ray is to grasp (*intelligere*) the spiritual sense of scripture through the practice of theology.[138] Here John obliquely compares his own understanding *(intellectus)* of future events to the academic understanding of scriptural truth and, of course, to Joachite exegesis. John engages a number of biblical and extra-scriptural sources throughout his writings, especially in the *Liber ostensor*, as a means of proving his divine insight. He is not, however, stating this his insight was gained by study—at least not solely by study. Rays four through six are fairly banal. Through the kind of preaching prescribed by Saint Francis the illumined brother will bring people to repentance.[139] Fifth, he can explain scripture to people (based on their level of ability), and if he has achieved the sixth ray of illumination, then he may write theological books.[140]

The seventh and highest ray, however, illustrates his Olivian-Joachite roots, as it is demonstrated by "interior profundity of all the prophetic senses of sacred scripture and of three general *tempora* of the world," which is to say the grasp of salvation history past, present, and future. John goes on here to call this interior profundity "the spirit of understanding (*spiritus intelligencie*)."[141] In this passage, John has set up Joachite exegesis as the apotheosis of academic theology, and implies that he has reached at least the final stage of illumination, far beyond what is necessary to be among the elect. As important, he has framed what to him seems an orthodox explanation for his power of foretelling.

As John continues to travel up the grades of sanctity, other elements of his own experience come to light. John calls the seventh grade "transcendence (*transcensio*)" or "ecstasy (*extazis*)."[142] Here we have left behind the academician and have met the mystic. Though John does not discuss the "heavenly sweetness" that overwhelmed him in his moments of understanding—discussing instead the edifying effects of transcendence—he lists prophecy among the outcomes of ecstasy.[143] This description of prophecy opens a door for John to both be and not be a prophet. He states that there are modes of prophecy that continue to operate "up to now . . . and without interruption to the end of the world."[144] Obviously, the prophecy of the Old Testament is complete, and latter day prophets neither predict the advent of Christ (although the second advent seems fair game) nor reveal additional divine law.[145] There are seven modes of prophecy, however, that will continue to be manifested until the end of time. John's description of these manners of prophesying is very close to the descriptions of his own experiences.

The first, and least, mode of prophecy is to see God or an angel in a dream.[146] John maintains he has had sleeping visions, though these are less important.[147] During his 1340 and 1344 visions he was awake, and an angel visited him while he was at Toulouse with a true foretelling of the future.[148] John reports that the angel visited him in the ninth hour (*hora nona*), which, if John was following liturgical time, would have placed the vision in the mid-afternoon, so John would probably place these important angelic communication as part of the sixth manner of prophecy which is "to hear the future while awake, from the mouth of the holy angels."[149]

Returning to John's list, the second manner of prophecy is Joachite exegesis. He calls out Joachim by name as an exemplar of understanding (*intelligenciam*) of the scriptures.[150] Undoubtedly John saw himself partaking of this tradition as well. Joachism is also the subject of the third and fourth modes of prophecy, which have to do with logically reasoning out future events from scripture.[151] The capstone of prophecy is also thoroughly Joachite: "to prophesy through the concord of both divine testaments."[152] Through this description, John locates his own prophetic activity as a high grade of divine activity. We can conclude that since prophesying is a mode of ecstasy, the "weight of celestial sweetness" felt when his understanding was opened was a concomitant aspect of this experience—integral rather than incidental.

We can now see John's claim not to be a prophet as the straw man it was meant to be. John made no claims to being a prophet like those of the Hebrew Bible, but such assertions were not meant to diminish his role as a predictor of the future. Rather they opened the door to a lengthy discussion of his gifts. Thus, John can claim without too much contradiction that he is not a prophet, while at the same time describe himself as the "new Hector" foretold by Joachim who would announce the doom of "new Troy—Avignon the sinner."[153]

It is no surprise why John paints himself as a new Hector in the *Liber ostensor*.[154] He must have been struck by the fact that Joachim spoke of a prophet who was believed to be *phantasticus* while speaking the truth about future events. This passage allows John the opportunity to turn his condemnation into proof of his veracity. In the course of demonstrating that the prophecy of the new Hector is about him, John makes full use of his persecutors' invective; he is thought "mad and delirious, thoroughly out of his mind and insane."[155] As importantly, in his most sustained narrative of his many imprisonments and tortures, he channels all of his sufferings for the truth of God into further evidence of the significance of his prophecy.[156] It should come as no surprise that the suffering he describes corresponds with the eighth grade of sanctity, *sacrifice* (*sacrificatio*) or martyrdom.[157] Though the seven species of martyrdom need not concern us here, suffice it to say that John enunciates many aspects of martyrdom that require suffering, but not death.[158]

What does concern us is that John offers in his section on martyrdom twelve rules (*regula*) for avoiding the snares of Antichrist. It is the first of these rules—a recapitulation of Olivi—that concerns us, as it has to do less with martyrdom than with the dangers inherent in visionary activity.[159] Hewing again to scriptural numbers, John tells us there are fourteen related parts to the rule. Some specific points shed light on John's case. First, John advises against cultivating a desire for visions, raptures, or revelations, since this is an opportunity for demonic deceit. Second, beware of pride in your prayers or meditation. One is also to avoid anyone who is mad as well as those who praise the delusions of the mad as true visions. One also needs to take great care when attending to visions and gauge whether they incline one for or against the teaching of Christ.[160] Finally, if through visions you are moved to undertake a pleasing task—say for instance, writing a book about the quintessence—you ought to pause to examine the circumstances of your vision and the probable impact of the work itself before continuing.[161]

Suffice it to say that John does not lack nerve, though the *Liber ostensor*, written some years after both the *Book of Hidden Events* and *De considerati-one*, exudes a certain confidence unmatched in those two earlier texts. The fact that John admits to no doubts in the *Liber ostensor* does not mean that he never suffered them. In the *Book of Hidden Events*, he tells his examiner to "judge whether it was almighty God himself who opened understanding for me . . . or whether I was deceived by an unclean spirit."[162] This was intended as a rhetorical question, no doubt, but in his later works John offers no such overt opportunities for ecclesiastical commentators to contradict him. We know that John was aware of the dangers of false prophecy—the danger if ecclesial officials believed him corrupted as well as the danger to his soul if he actually was.

In the course of proving his status as the new Hector in the *Liber ostensor*, John indulges in a confession about his own complexion.[163] "Know that almost since I was at my mother's breast, I sensed in myself from my own evil ways a destructive complexion . . . since I was—and still am—very prone from excess pride and arrogance (*inflatio cordis*) to anger and moved to raging over nothing, so that I curse and demean in others the things which I do not know how to check in myself."[164] To the extent that God has lifted this burden from John, he continues, "I abhor myself, seeing myself plainly as a beast without reason, since I see myself angered without cause or reason and everywhere swearing on the saints unnecessarily."[165] John is describing the classic traits of a choleric, especially overwhelming pride and anger.

"An abundance of choler," John tells us in *De consideratione*, is not only a cause of fevers, but "is commonly linked to alienation of sense[s] and the presence of fantastic things."[166] Of course, people of all humoral dispositions may have false visions, and cholerics do not seem to be deviled by the worst of these, a dubious distinction that belongs instead to melancholics.[167] Still, the amount of time John spends in the *De consideratione* discussing cures for false visions and the attention he gives to visions in his own work suggest that John indeed had considered the possibility he might have been influenced by a poor humoral imbalance or demonic interference.

It is highly probable that John considered the quintessence a safeguard of the validity of his own visionary experiences as well as those of future evangelical men. John mentions two different instances of having used the quintessence to heal himself, though in each case for physical ailments rather than for madness.[168] John's use of the quintessence may have aided him in determining that his visionary experience was not just profound, but divine. As a guarantor of physical and mental health, the quintessence seemed a more certain manner of knowing the truth about visionary experience than interpreting the nature of visions themselves. Whether or not John intended to create a cure for (his own) madness, the development of the elixir afforded him a unique means of reclaiming his self-determination. John's proto-science became the guarantor of his religious experience and the proof of his chosen status. No amount of spiritual reflection or mystical experience could have done the same.

The subjunctive world, whether conjured by ritual or not, often exists as a temporary reprieve, an impermanent repair of a disordered and fragmented world. Seligman et al. understand the impermanence of the ritually achieved subjunctive world to be "tragic," always returning to its broken state.[169] The tragic is always just at the edge of John's consciousness. In the course of the *Liber ostensor*, John offers a prophecy of Joachim of Fiore, though he demurs at first to explain it:

> The ancient philosopher Anaxagoras prophesied by speaking plainly. Hector after his death denounced the destruction of Troy openly; yet while he was sleeping in his bed he foresaw it. Thus a new Hector, a warrior in the new church, shall announce a new destruction of Troy, but he shall be thought mad (*fantasticum*).[170]

Later, he concedes this point, stating, "I have not explained this prophecy of Joachim *that was made about me* [emphasis mine].[171] In the subjunctive world, every accusation of madness is a proof of John's sanctity. The world as it could be was the world where John's imprisonment, suffering, and shaming made sense and were given meaning. Much as Francis's liturgy drew up a subjunctive world where God's justice ultimately remedied earthly privations, the power of the quintessence put the lie to allegations of madness and proved John's soul was as it should be, even if only to John himself.

* * *

This chapter began with the claim that medieval alchemy was not magical. My argument for alchemy's subjunctivity makes this no less true. It even reinforces this medieval claim. The difference between alchemy and magic is not, however, due to its scientific character, at least not in full. Rather, Bacon and Rupescissa—and even Vitalis, though far less dramatically—set alchemy within the confines of the sacral cosmos—a subjunctive universe where God's plan and laws could be realized within the natural world. Within the Christian sacral cosmos, magic was reserved for frauds and deceits. The elixir was instead a manifestation of how God's power and grace permeated the world He created.

The subjunctive world of medieval Franciscans, however, was quite different from that of the Hermetic authors whose ideas and works Western inheritors appropriated. There is no singular subjunctive reality. Rather it is a manifestation of how ritual actors understood the world should be. Therefore Franciscan authors considered themselves neither magicians nor theurgists. Their subjunctive reality was not one where individuals coerced divine or demonic powers to act on their behalf, but rather a world where God had mercifully disposed to His people the possibility of healing their bodies and even transforming their earthly existence. Therefore, approaching the creation of the elixir as an act of devotion demonstrated humility, gratitude, and subservience.

While the subjunctive is most clearly defined in the ritual mode, the attempts to render the subjunctive world into discourse are a means of capitalizing on the connections and meaning made in ritual activity. As important, the distinction between the subjunctive world and day-to-day existence was also clear to these friars. Theirs was not only a cosmos ruled by an omnipotent, omniscient, and yet present God. Their experiences also manifested the disjunctions between the world of what should be and what was. John suffered from privation and imprisonment. Bacon could not win needed support for his agenda. Most importantly, both Bacon and Rupescissa seemed to have some experience of the elixir, but at no time did the elixir actually demonstrate the fullness of its potential. The subjunctive world was no less real, however. There was a very strong sense among the authors that the elixir *could be* manifested empirically and maybe even that it had been, but that its creation was contingent on the disposition not just of those who made it, but also of God. Realizing the subjunctive, through rituals and through the production of the elixir, made sense not only of the world, but also of the alchemist.

Conclusion

At one point in the *De consideratione*, John invokes God as "Most blessed God, creator of secrets."[1] Both Bacon and Rupescissa often employed the term "secret" in their work. For Bacon the term derived from pseudo-Aristotle's *Secret of Secrets*, a major influence on his alchemical and astrological theories. Rupescissa also used the term "secret," but did so more generally—not just for the hidden properties of alchemical ingredients and compounds, but also for prophecy. A secret in the medieval, as in the early modern, period, was not a strictly naturalist *nomen*.[2] It could include formulae and even recipes for practical jokes, but was synonymous with both *experimenta* (observations) and *mirabilia* (wonders).[3]

Thus a secret was something that one uncovered not through reason, but via experience, be it trial and error or, at least for the medieval examples, revelation. The secrets traditions of early modernity prized revealed knowledge about alchemy, but also domesticated, commodified, and demystified alchemical products and practices, suggesting that shared terminology may obscure some significant differences between the two periods. William Eamon signals as much in his discussion of the decline of pseudo-Aristotle's *Secret of Secrets* in favor of a new, more popular, and more practical secrets literature such as *The Secrets of Alessio Piemonte*.[4] In the medieval period, as Steven J. Williams has pointed out, the secrets tradition revolved around the university. The concentration of scholars at Paris briefly elevated the place of *The Secret of Secrets* and saw it transmitted to other academic milieux.[5] Concentration in academic circles, rather than diffusion among secular readers, had it drawbacks, however. *The Secret of Secrets*' place among the corpus of Aristotelian works was rather limited, especially compared to logic and the *libri naturales*. Eventually many if not most medieval scholars also recognized it for a pseudonymous work. Moreover, as Eamon has pointed out, *The Secret of Secrets* was more esoteric than empirical—just the kind of text early modern secrets literature abhorred.[6]

One of the more interesting issues emerging from the focus on empirical secrets in the early modern period is Eamon's question whether "technical literacy contribute[d] in some measure to the 'disenchantment of the world' that Max Weber notes was one of the hallmarks of the modern culture."[7] While this book makes no claims toward elucidating the early modern period, Eamon's query does bear on the elixir. Eamon is, of course, speaking about a much more diffuse genre of secrets books and recipe collections in early modernity, while here I am tackling only a few texts, none of which were standard titles in a medieval library. Still, in a micro sense, each alchemist had to come to grips with the empirical result of his actions. Did the failure to produce consistently an elixir that precluded aging, healed wounds, and the like affect the disposition of these Franciscans toward their alchemical praxis? Clearly it did not. One can, of course, make all kinds of excuses for misinterpretation of empirical evidence. *Maybe* John of Rupescissa took the elixir at the same time his chest cold was naturally clearing up. Distilled alcohol does have an antiseptic effect and could have cured some skin ailments. These kinds of speculations are quite beside the point, however. There is no getting around the fact that John's elixir did not do all the things he claimed it could. Nor did Bacon ever produce an elixir that would transform people into Christians. Even the somewhat less hyperbolic claims made in *Pro conservanda* could not have survived any sort of rigid empirical testing in regard to its effect on snakebites or other calamitous injuries. This is not to pass judgment on Eamon's question. Rather, this is a case where periodization matters.

The medieval world of Bacon, Vitalis, and John of Rupescissa was not awash in books of secrets or alchemical texts. While the study of alchemy might have taken place around universities, it was only very briefly a staple of the curriculum during the thirteenth century. The kind of widespread secrets discourse among the urban laity that existed after the birth of the printing press was quite different from the medieval intellectual environment. There is a transition of sorts, on display in the *De chymico miraculo*, ascribed to Bernard Trevisan that was circulating no later than the second half of the fifteenth century and found a life in print in 1583.[8] While the author of this text prays to God for aid and believes that the truth of alchemy is evident in the Book of Genesis, he objects to Rupescissa's alchemy (and other medieval alchemies) on empirical grounds.[9] The directions Rupescissa supplies do not produce the hoped-for effects, and alchemical knowledge is gained through trial and error. There is something, to be sure, reminiscent of earlier alchemical texts that criticized the work of rivals, but the crux of the author's complaint with Rupescissa is based on an inability to replicate results. This critique comes more than a century after Rupescissa died, however, during a period when the definition of secrets appears to be changing. In the early modern period, especially in the Italian context, the term "secret" was nearly synonymous with "recipe (*ricette*)." This was not the case everywhere, but the relationship between the two ideas was fluid and suggested that most early modern secrets could be treated as recipes, even if that was not always the intent of the author. The contigent elements of Rupescissa's subjunctive elixir seem entirely lost on the author of *De chymico miraculo.* In addition to chronological difference, the social context of early modern secret readers was quite different. While early modern readers of secrets books may have been devout, the religious praxis and daily life of a medieval Franciscan was quite distinct from that of a middle-class physician in Renaissance Florence or an educated matron of seventeenth-century Flanders. Therefore a counterpoint to Eamon's question is whether subjunctivity could have held up as a strategy in the face of such a widespread diffusion of alchemical practice. Bacon, John, and Vitalis could have excused themselves if an elixir did not produce the desired effect. The stars might have been wrong, the distillation incomplete, or even that God did not desire the elixir to work in a certain instance. Since most of their experience with the elixir was, as their texts indicate, theoretical, it was probably more important it could work than if it actually did.

There is a sense that Bacon and Rupescissa were working in a closed system. The structure of this book reflects that. The sacred, created world discussed in the first chapter is, essentially, the subjunctive world of the elixir. The cosmological frame proffered by Christianity was essential to the concept of the elixir's characteristics as well as to the possibility of its existence. The shared cosmology, however, seems to have had less impact than the rather idiosyncratic way in which each approached or interpreted the relationship of alchemy to Christian cosmology. For Bacon and Rupescissa, the elixir was not only a cure, but a tool for realizing eschatological ends. The pursuit of the elixir was defined as an essentially Christian practice that served a role in salvation history. Vitalis, on the other hand, kept alchemy within the framework of Christianity, but did not describe it as a Christian practice per se. While the constraint of the encyclopedic genre certainly played a major role in limiting Vitalis's comments on the elixir, it also seems to be the case that Bacon's and Rupescissa's elixirs *required* theorization precisely because they were located at the heart of a forthcoming apocalyptic drama. If alchemy was a Christian practice, then it also *had* to work. There was urgency to realizing the elixir, subjunctively or otherwise, that is simply not present in Vitalis's text. Therefore, as much as I have described religion as a constructive aspect of Bacon's and Rupescissa's alchemy, furnishing it with both purpose and models of alchemical effects, it also may have felt coercive. For Bacon and Rupescissa, to call into question the efficacy of the elixir was to tacitly call into question what they understood to be Christian truth. At the same time, as much as Bacon's and Rupescissa's elixirs hinged on their understanding of Christian truth, their ideas about both the elixir and Christian truth evolved over time. Bacon's inestimable glory in its early phase is basically a theriacal medicine. Over time Bacon imbues it with more potency. John of Rupescissa implicitly refutes the subject of his first alchemical work, the *Liber lucis*, a treatise on making gold, when in *De consideratione* he condemns metallurgical alchemy. Thus, the constraints of belief placed on John and Roger were largely of their own making, and represent a moment in their intellectual chronologies.

It may also be tempting to conclude that for science to emerge as a rational discipline, shedding subjunctivity was critical. I would argue, however, that such a conclusion is quite incorrect. Subjunctivity was not, and is not, antithetical to scientific pursuit. Subjunctivity potentially embodies the imaginative process of science. It assumes meaning and order. The ability to ideate beyond the constraints of one's immediate environment and experience is crucial to forming theoretical constructs and hypotheses, and taking the kinds of imaginative leaps that characterize everything we celebrate about science. Bacon's and Rupescissa's reliance on the subjunctive mode is not inimical to science. It is part of it. Their subjunctive truths emerged, however, in a cultural matrix quite alien to our own, in spite of our general sense that we exist in continuity with Western traditions.

As much as these authors adapted different alchemical traditions and practices in their discussion of the elixir, they did something similar with Franciscan and Christian tropes and truth claims. Christianity's impact on these authors is unquestionable, but not monolithic. There was no singular "Christian" manner of producing the elixir, or performing alchemy. Alchemical and Christian discourses did not exist in a dialectical relationship, but they were in conversation. This dialogue was mediated by authors whose particular experience with both led to distinctive formulations. In this sense, there is no religion qua religion or science qua science. There are only elixirs, whose compositions, effects, and modes of production derive from the store of intellectual material available to their authors and the power of their individual imaginations.

NOTES

INTRODUCTION

1. Crisciani and Pereira, *L'Arte del Sole e della Luna*, 8–13.

2. (pseudo?) Michael Scot, *Ars alchemie*, in Thomson, "The Texts of Michael Scot's Ars Alchemie," 344, and the corresponding transcription, Haskins, "The 'Alchemy' Ascribed to Michael Scot," 354, 355.

3. Bartlett, *The Natural and Supernatural*, 122.

4. Pereira, "Alchemy and the Use of Vernacular," 336; Crisciani, "Artefici *sensati*," 135–36.

5. Crisciani and Pereira, *L'Arte del Sole e della Luna*, 3; Pereira, "Alchemy and the Use of Vernacular," 336; Ingham, *The Medieval New*, 28.

6. The terms *Hermetic* and *Hermetica* present some particularly thorny problems. The adjective *Hermetic* is often employed to describe anything secret or alchemical, and sometimes a corpus of works or a philosophical doctrine. Following Kevin van Bladel, I will employ the term *Hermetica* to mean works attributed to Hermes, and the adjective *Hermetic* to that particular body of texts. By doing this I hope to avoid association with the alchemy and alchemists of the early modern period, who embraced various notions under the rubric of hermetism or hermeticism, and forgo the assumption of hermeticism as a discrete category in the Late Middle Ages. van Bladel, *The Arabic Hermes*, 17–21. Crisciani and Pereira, *L'Arte del Sole e della Luna*, 7–8.

7. Crisciani and Pereira, *L'Arte del Sole e della Luna*, 13.

8. McVaugh, "The 'Experience-Based Medicine' of the Thirteenth Century," 111–12.

9. Power, "A Mirror for Every Age," 657–92, at 675. Fascination with the historiography of Bacon is shared by Patricia Ingham. See Ingham, *The Medieval New*, 48–72.

10. For discussion of Bacon's biography see Jeremiah Hackett, "Roger Bacon: His Life, Career, and Works," 9–20. For a discussion of the various theories on Bacon's actual date of birth, see Hackett, "Roger Bacon," 90–102.

11. Glorieux, *Répertoire des maîtres en théologie*, II, 137.

12. Marrone, *In the Light of Thy Countenance*, II, 267. Cf. Glorieux, *Répertoire des maîtres en théologie*, 137.

13. For a description of John's life, see Lerner, "Historical Introduction," 15–29, 83–85; Lerner, *The Feast of St. Abraham*, 74–78; Bignami-Odier, *Etudes*, and Vauchez, "Introduction," in Rupescissa, *Liber ostensor*, 12–62.

14. Smith, "Secrets and Craft Knowledge," 52.

15. A fine introduction to Franciscan alchemy and its philosophical genealogy can be found in Pereira, "I francescani e l'alchimia," 117–57. I am dealing only with descriptions of the elixir that can be reliably assigned to Franciscan authorship. This therefore excludes the pseudo-Lullian corpus. While Ramon Lull (d. 1316) was a Franciscan tertiary, he was not the author of the many alchemical works that bear his name. See Pereira, *The Alchemical Corpus Attributed to Raimond Lull.*

16. Bynum, *Christian Materiality*, 18, 285.

17. DeVun, *Prophecy, Alchemy*. Paravicini Bagliani, "Ruggero Bacone e l'alchimia della lunga vita."

18. Lawrence Principe has made a cogent argument for the need to examine alchemy contextually, rather than teleologically. Principe, "Alchemy Restored," 305–12.

19. Chakrabarty developed this theory in response to another totalizing discourse: capitalism. Chakrabarty, *Provincializing Europe*, 48.

20. Newman did not see the religious turn as somehow precluding naturalistic alchemical texts. Newman, *Promethean Ambitions*, 82–83. Also useful is the chronology presented by Michela Pereira, who cites the *Liber secretorum alchimie* (1257) as the first Western alchemical text to engage in cosmological speculation. Pereira, "Cosmologie alchemiche," 363–410, at 388.

21. Matus, "Alchemy and Christianity in the Middle Ages," 934.

22. Eliade, "Alchemy," I, 234.

23. Eliade, *The Forge and the Crucible*, 8. See discussion of this approach in Principe and Newman, "Some Problems with the Historiography of Alchemy," 408–11, and Matus, "Alchemy and Christianity in the Middle Ages," 935–36.

24. Halleux, *Les textes alchimiques*, 142.

25. Obrist, "Alchimie et allégorie scripturaire au Moyen Age," 137; DeVun, *Prophecy, Alchemy*, 102–3.

26. See, for example, Salimbene, *Cronica*, 331–32.

27. Thorndike, *A History of Magic and Experimental Science*, III, 347.

CHAPTER I. FRANCISCANS AND THE SACRAL COSMOS (THE CONTEXT OF FRANCISCAN ALCHEMY)

1. See Introduction.

2. Hugh of St. Victor, *Didascalicon*, I.10.

3. "*De natura vero corporea quantum ad fieri haec specialiter tenenda sunt, quod sex diebus sit in esse producta, ita quod in principio, ante omnem diem creavit Deus caelum et terram.*" I have used the Quaracchi edition of the *Breviloquium* (full reference in the

Bibliography), but given that many texts have adopted the Quaracchi text yet with different pagination, further reference will include the Book and Section of the work that are found generically. Bonaventure, *Breviloquium*, II.2.i.

4. The bibliography on this subject is vast. See especially Lambert, *Franciscan Poverty*, and Burr, *The Spiritual Franciscans*.

5. Francis of Assisi, "Letter to All the Faithful," in Habig, *Omnibus*, 95.

6. For instance, Francis's testament, often used to demonstrate Francis's unhappiness with the state of his Order just before his death, considers reverence for the Eucharist "above all else" and enjoins the friars to keep it in "richly ornamented" places. This sacrament also receives pride of place in his *Admonitions*. See Habig, *Omnibus*, 6, 77–79.

7. Martel, *Les cathares et l'histoire*; Moore, *The War on Heresy*, 332–41.

8. The Cathars did not constitute, as once was believed, an organized "counter-Church," and they held, as did normative Christians of the era, a variety of doctrines and ideas that were not necessarily reconcilable with one another. Cathar texts demonstrate at least two different "dualisms"—mitigated as well as absolute. Inquisition records reveal further diversity of views, though all perceived the created world negatively. Lambert, *The Cathars*; René Nelli, *Écritures cathares*; Brenon, *Les femmes cathares*; cf. Moore, *The War on Heresy*. Regarding the strict dietary restrictions of the Cathars, it is worth noting that Francis took the opposite tack and allowed his brethren (and himself) to eat meat, which was (in theory at least) a rarity in other monastic institutions of his day. Thompson, *Francis of Assisi*, 56.

9. Vauchez, *Francis of Assisi*, 282. A good discussion of the general surge in and emphasis on preaching in the thirteenth century is found in Rouse and Rouse, *Preaching, Florilegia, and Sermons*, 43–64, esp. 59. For a more recent discussion of the Franciscan part in the upswell in preaching at the beginning of the thirteenth century, see Johnson, "The Franciscan Fascination with the Word," 2–5.

10. Sorrell, *St. Francis and Nature*, 98–105. Moloney, *"Canticle of Brother Sun" Reassessed*, 58. Nairn, "St. Francis of Assisi's *Canticle*," 204. Thompson, *Francis of Assisi*, 124.

11. Thompson, *Francis of Assisi*, 124–25.

12. Both the original Umbrian and the English translation are from Moloney, *"Canticle of Brother Sun" Reassessed*, xxi–xxiv.

13. Sorrell, *St. Francis and Nature*, 115.

14. Moloney, *"Canticle" of Brother Sun Reassessed*, 44.

15. Thompson, *Francis of Assisi*, 124. Nairn, "St. Francis of Assisi's *Canticle of Creatures*," 203, 209–10. Vauchez, *Francis of Assisi*, 279–80.

16. Thompson, *Francis of Assisi*, 122.

17. Blastic, "Preaching in the Early Franciscan Movement," 38. Vauchez, *Francis of Assisi*, 281.

18. Sorrell, *St. Francis and Nature*, 87–92. Some of the better discussions of Sorrell's work and the problem of mysticism are found in Moloney, *"Canticle" of Brother Sun Reassessed*, 71–85; Nairn, "St. Francis of Assis's *Canticle of Creatures*," 206; Vauchez, *Francis of Assisi*, 273–74; Thompson, *Francis of Assisi*, 56.

19. This paragraph follows Moloney, *"Canticle" of Brother Sun Reassessed*, 46–52.

20. Thomas of Celano, "The Life of St. Francis," in Armstrong, Hellman, and Short, *Francis of Assisi: Early Documents*, 284 quoted in Moloney, *"Canticle" of Brother Sun Reassessed*, 79.

21. Vauchez, *Francis of Assisi*, 279.

22. Bonaventure, *Itinerarium*, 7.2–3.

23. "*Significant autem huiusmodi creaturae huius mundi sensibilis invisibilia Dei, partim quia Deus est omnis creaturae origo, exemplar et finis.*" Bonaventure, *Itinerarium* 2.12.

24. Thompson, *Francis of Assisi*, 54.

25. cf. Nairn, "St. Francis of Assisi's *Canticle of Creatures*," 206–7.

26. See Chapter 2 for a discussion of the fifth element.

27. Nairn, "St. Francis of Assis's *Canticle of Creatures*," 206–7.

28. Paolo Capitanucci argues that the alchemical activities of the Franciscan Bonaventure of Iseo may have become conflated with the legend of Elias. See Capitanucci, "Francescani e alchimia fra mito e realtà," 164–65.

29. Bonaventure, *De Reductione Artium ad Theologiam*, 42–43.

30. See below.

31. Thompson, *Francis of Assisi*, 55–56.

32. Şenocak, *The Poor and the Perfect*, 97–104.

33. Julian of Speyer, *Vita Sancti Francisci*, 356; English in Armstrong, Hellman, and Short, *Francis of Assisi: Early Documents* I, 400–401, quoted from Şenocak, *The Poor and the Perfect*, 104.

34. Thompson, *Francis of Assisi*, 224.

35. Vauchez, *Francis of Assisi*, 200.

36. Abelard, *Expositio*, 7.

37. Abelard, *Expositio*, 14–16, 25–26.

38. Abelard, *Expositio*, 41–51.

39. Abelard, *Expositio*, 51.

40. Pereira, "Cosmologie alchemiche," 363–410, at 395–97.

41. Bonaventure, *Breviloquium*, Prol., 1.

42. Monti, "Introduction," xiv–xxxviii, at xxi.

43. Bonaventure, *Breviloquium*, Prol., I, 201–2.

44. Bonaventure, *Breviloquium*, II, 2, i–iii.

45. "*creatura mundi est quasi quidam liber, in quo reluceret, repraesentur et legitur Trinitas fabricatrix.*" Bonaventure, *Breviloquium*, II.12.i.

46. Solignac, *La théologie symbolique*, 26.

47. Bonaventure, *Breviloquium*, II.3. i–ii.

48. Bonaventure, *Breviloquium*, II.3.v., II.4.i.

49. "*productionem rerum generabilum et corruptibilem, scilicet mineralium, vegetabilium, sensibilium et corporum humanorum.*" Bonaventure, *Breviloquium*, II.4.i.

50. "*distinctionem orbium nec caelestium nec elementarium.*" Bonaventure, *Breviloquium*, II.5.i.

51. Bonaventure, *Breviloquium*, II.5.ii–x.

52. Bonaventure, *Breviloquium*, II.11.vi.

53. Solignac, *La theologie symbolique*, 33–34.

54. Bonaventure, *Breviloquium*, II.12.v.

55. Bonaventure, *Breviloquium*, II..12.ii–v.

56. Solignac, *La théologie symbolique*, 37.

57. Anderson, *A Call to Piety*, xi.

58. Olivi, *Olivi on Genesis*, 1.

59. Olivi, *Olivi on Genesis*, 1.

60. "*Sub hac autem arbore tota Dei trinitas et trina angelorum hierarchia et tria virorum genera exsistentium sub lege naturae, Scripturae, et gratiae*," Olivi, *Olivi on Genesis*, 1.

61. Burr, "The Antichrist and the Jews," 33–34.

62. "Therefore in all of sacred Scripture, the mystery of the Trinity is the first to appear, since just as the Son arises from the Father as well as does the Holy Spirit from each, so does the letter of the New Testament arise from the letter of the Old Testament as well as from special understanding. (*In tota igitur Scriptura sacra primo occurrit Trinitatis mysterium, quia sicut a Patre oritur Filius et ab utroque Spiritus Sanctus, sic de littera Veteris Testamenti oritur littera Novi Testamenti et ex utraque specialis intelligentia utriusque.*)" Olivi, *Olivi on Genesis*, 7.

63. Olivi, *Olivi on Genesis*, 6–7.

64. "*Quidam tamen putaverunt quod lux illa a loco suo non mota, postquam per spatium diei usualis illuminaverat orbem, retrahebat radios suos, et sic fiebat nox in toto orbe*," Olivi, *Olivi on Genesis*, 56.

65. For a general sense of Augustine's opinions on Genesis, see Hill, "General Introduction," 13–22. Citations of Zycher's Latin edition of the text will be referred to as Augustine, *De Genesi ad litteram.* English translations come from Hill's edition, Augustine, "The Literal Meaning of Genesis."

66. Augustine, *De Genesi ad litteram,* 8. Augustine, "The Literal Meaning of Genesis," 173. Understanding "the beginning" as the Son is hardly a stretch for a Christian exegete. The Gospel of John opens, echoing Genesis, "*In principio erat verbum et verbum erat apud Deum et deus erat verbum*" (Jn 1:1) (The first verse of John is identical in both the Latin Vulgate and extant fragments of the *Vetus Latina*, the pre-Vulgate Latin translation of the scriptures from which Augustine worked.) Burton, Balserak, Houghton, and Parker, "Vetus Latina Iohannes Synopsis 2.0."

67. Augustine, *De Genesi ad litteram,* 9. Augustine, "The Literal Meaning of Genesis," 173.

68. Augustine, *De Genesi ad litteram,* 10. Augustine, "The Literal Meaning of Genesis," 173.

69. Olivi, *Olivi on Genesis*, 9.

70. "*Litteralis quidem non est, quia tota littera a prima die usque ad septimam ostendit aperte se loqui de diebus et noctibus corporalibus*," Olivi, *Olivi on Genesis*, 43.

71. Olivi, *Olivi on Genesis*, 48–50.

72. Olivi, *Olivi on Genesis*, 52–53.

73. Olivi, *Olivi on Genesis*, 29–31.

74. Olivi, *Olivi on Genesis*, 65–66.

75. For details on Nicholas's career, see Delmas, "Nicolas de Lyre Franciscain," 18–21.

76. Dahan, "Herméneutique et méthodes d'exégèse," 111–12.

77. Nicholas of Lyra, *Postilla*, c1R, col. i. For Averroës's more complete (and somewhat more complicated) view, see Hyman, *Averroës'* De substantia orbis, 30–31, 53n.

78. Nicholas was certainly familiar with Joachim, but his reading of the key Joachite book of the Bible, Revelation, was quite distinct from the way Joachim and his Franciscan follower Peter Olivi interpreted the text. See Krey, "The Apocalypse Commentary of 1329," 267–69. Gilbert Dahan prefers the term "tension" to describe the relationship between Aristotle and the biblical description of the Hexaemeron, Dahan, "Herméneutique et méthodes d'exégèse," 112–13. On Nicholas's commentary on Genesis as Christocentric, see Patton, "Creation, Fall and Salvation: Lyra's Commentary on Genesis 1–3," 21.

79. Pereira, "I francescani e l'alchimia," 117–57.

80. I take up this issue in more detail in Chapter 3.

CHAPTER 2. THREE ELIXIRS

1. Principe, *The Secrets of Alchemy*, 39.

2. Grant, *Planets, Stars, and Orbs*, 577–78. Karpenko, "Alchemical View of the Origin of Metals and Transmutation," 19–20.

3. Principe, *The Secrets of Alchemy*, 38–39.

4. Halleux, *Les Textes alchimiques*, 70. Crisciani and Pereira, *L'Arte del Sole e della Luna*, 7. Newman, *Promethean Ambitions*, 43.

5. Pereira, "Alchemy and the Use of Vernacular," 336.

6. Avicenna, *Avicennae De congelatione et conglutinatione lapidum.* The addition was generally accepted as the work of the Philosopher himself, a final book of *Meteorology. De congelatione* is a fascinating addition since, while it attacked the possibility of alchemical transmutation, it also transmitted under Aristotle's name a considerable amount of metallurgical theory and cemented the discipline as worthy of attention by readers of Aristotle. Newman, *Promethean Ambitions*, 43.

7. See Barbara Obrist's introduction to Constantine of Pisa, *The Book of the Secrets of Alchemy*, 3–5.

8. See Paul of Taranto, *Theorica et Practica.*

9. Constantine of Pisa, *The Book of the Secrets of Alchemy*, 70, 98–101. Also see Obrist's discussion at 32–35 of that volume.

10. Also circulating was the theory that metals found in the earth were at various stages of maturation, with gold most mature. Grant, *Planets, Stars, and Orbs*, 577–78.

11. Roger Bacon, *Part of the Opus tertium of Roger Bacon*, ed. A. G. Little (Aberdeen: University Press, 1912), 82; cited henceforth as Little, *Opus tertium*. Little's publication complements what is found in Roger Bacon, *Opus tertium* in *Opera quaedam hactenus inedita*, ed. J. S. Brewer (London: Longman, Green, Longman, and Roberts, 1859), cited henceforth in the notes as Brewer, *Opus tertium*.

12. Little, *Opus tertium*, 77, 81–82.

13. For a sense of the chronology of Bacon's works, see Williams, "Roger Bacon and the Secret of Secrets," 366–72.

14. Constantine of Pisa, *The Book of the Secrets of Alchemy*, 104–5. Roger Bacon, *The Opus majus of Roger Bacon*, ed. J. H. Bridges (Oxford: Clarendon, 1897), 35; cited henceforth as Bridges, *Opus maius*.

15. Little, *Opus tertium*, 86.

16. Matus, "Resurrected Bodies and Roger Bacon's Elixir," 326–27.

17. "*Generatio enim hominum, et brutorum, et vegetabilium, est ex elementis et humoribus, et communicat cum generatione inanimatarum rerum.*" Brewer, *Opus tertium*, 39.

18. Newman, "An Overview of Roger Bacon's Alchemy," 319–23. Newman has shown that Bacon's humoral-elemental theories rely on pseudo-Razi's *Lumen luminum* (*Light of Lights*) and pseudo-Avicenna's *Liber de anima*.

19. Bacon, *Opus minus*, 364.

20. Newman, "The Philosopher's Egg," 88–89.

21. Philosophical agriculture is also necessary. See Bacon's prologue to *On the Multiplication of Species*, in "Appendix," Bacon, *De multiplicatione specierum*, 349. According to Lindberg, the introductory piece Bacon wrote for the work is probably much later than the work itself. *Multiplication* dates to the 1240s or 1250s, and the prologue was added in the 1260s.

22. Hackett, "Roger Bacon on *Scientia experimentalis*," 294–95.

23. Little, *Opus tertium*, 51–52.

24. Roger Bacon, *Frater Rogerus Bacon in libro sex scientiarum*, 185. (Though many of the texts in Little's collection are not the work of Bacon, William Newman has verified the authenticity of the *Book of Six Sciences*.) Later Bacon links this sort of foresight to pseudo-Ptolemy's (Abu Jafar Ahmed ibn Yusuf, d. 912) term *cognitionem ex se* from the *Centiloquium*, which Hackett has characterized as "instant intuition or inspiration." Hackett, "Roger Bacon on *Scientia experimentalis*," 295–96.

25. Little, *Opus tertium*, 18.

26. Kieckhefer, *Magic in the Middle Ages*, 12.

27. Bailey, *Magic and Superstition*, 2.

28. Bailey, *Magic and Superstition*, 3.

29. "*De istis scienciis naturalibus que vocari possunt inproprie geomancia, ydromancia, aerimancia, piromancia, que sunt vere partes philosophie, intendit Aristotiles in hoc libro, set translator non habuit in Latino nomina propria istis scienciis, ideo accepit nomina scienciarum magicarum que sunt similes aliquibus veris scienciis.*" Bacon, *Tractatus brevis*, 12.

30. Burke, *Opus maius*, 587.

31. Little, *Opus tertium*, 49.

32. Burke, *Opus maius*, 411. For some context on Bacon's view of old women, see Agrimi and Crisciani, "Savoir médical et anthropologie religieuse," 1281–1308.

33. Bacon, *Tractatus brevis*, 4–5.

34. Bacon, *Tractatus brevis*, 4–5.

35. Bacon, *Liber sex scientarum*, 181.

36. Bacon, *Liber sex scientiarum*, 182.

37. Burke, *Opus maius*, 625.

38. Serpents generally, and their "cousins" dragons more rarely, were topics of interest in medical writings of this era, due to the belief that illness was a type of poisoning, against which a physician could employ a universal antivenom, the *theriac* (*tyriac*). Getz, "Roger Bacon and Medicine," 359. Bacon's goal of purifying the body of corrupted elements operates from a similar paradigm, which he describes in his *On the errors of physicians.* See Newman, "An Overview of Roger Bacon's Alchemy," 324. Bacon also wrote a book on the subject of the *theriac*, the *Antidotarius*.

39. Burke, *Opus maius*, 625.

40. Burke, *Opus maius*, 624.

41. Burke, *Opus maius*, 624.

42. Burke, *Opus maius*, 624.

43. Bacon, *Liber sex scientiarum*, 184. Bacon goes on to say that God will raise such bodies from the ashes for both the righteous and the wicked, but intriguingly offers no gloss on the post-resurrection body of Christ. It is hard to know why he did not, though perhaps alchemical speculation on the nature of the Godhead wandered into territory that was too risky even for Bacon. Nevertheless, there would be differences between the bodies. For one, Christ eats after his resurrection (Luke 21: 41–43), and still carries the wounds of his crucifixion (John 20: 20, 27–29). Bacon's post-resurrection bodies require no nourishment.

44. Burke, *Opus maius*, 624.

45. Peter Lombard, *Sententiae*, 4, dist. LXIII–L.

46. "*ad terminos quod deus constituit et natura.*" Bacon, *Liber sex scientiarum*, 181–82. Bacon repeats this so often it is a stock phrase.

47. I discuss these issues in much more detail in Matus, "Resurrected Bodies and Roger Bacon's Elixir" 334–38.

48. Bynum, *Resurrection*, 235.

49. Bonaventure, *Commentary on the Sentences*, vol. 6, Lib. IV, Dist. XLIX, Part. II, Art. 1, Quaest. i, 582.

50. Bynum, *Resurrection*, 236, 250.

51. Bacon, *Opus minus*, 371–72.

52. Bynum, *Resurrection*, 200.

53. Newman, "An Overview of Roger Bacon's Alchemy," 331. Blood is a vehicle for extracting the primary humors. Brewer, *Opus tertium*, 86–87.

54. Bacon, *Tractatus brevis*, 8.

55. Galen himself was a champion of astrological medicine. Jacquart, "Le soleil, la lune et les états du corps humain," 239–56. See also Grant, *Planets, Stars, and Orbs*, 577–78.

56. "*meliorare complexiones eorum ut inclinentur ad bona et utilia sibi et aliis, tam in sapiencia quam in moralibus.*" Bacon, *Tractatus brevis*, 9.

57. "*Et si hoc maximum potest fieri, constat tunc omnia alia possibilia, scilicet ut homo deueniat ad magnam prudentiam et sapientam perfectam ut sciat se et alios regere, dei gratia adiutrice.*" Bacon, *Liber sex scientarium*, 185.

58. Little, *Opus tertium*, 53–54. Steven J. Williams has pointed out that Bacon likely misread the section of the *Secretum secretorum* where the secret of complexions is discussed, accounting for the idiosyncrasy of Bacon's notion. Williams, "Roger Bacon and the *Secret of Secrets*," 388.

59. Little, *Opus tertium*, 9–12. This is an ancient idea found in the Hippocratic corpus that remained influential past the Middle Ages.

60. "*Et docuit eum opera quibus alteraret regiones, et civitates infortunaret, et infatuaret eas, ut se juvare non possent. Et tunc regiones male complexionis alteravit in bonam, ut homines malarum complexionum reduceret ad bonas; quatenus per consequens reduceret eos ad bonos mores et ad honestas consuetudines, et sic permisit homines vivere, et tamen subjectos. Unde Aristoteles sic dixit ei: Altera aerem hominum malarum complexionum et permitte eos vivere. Nam aere alterato, alteratur complexio, et ad alterationem complexionum sequitur alteratio morum. Et hec fuit sapientia ineffabilis.*" Little, *Opus tertium*, 53–54.

61. "*alterantur homines in corpore et anima, ut in eis compleatur naturalis bonitas longeuitas, morum, prudentie et sapientie. . . . Et hoc est secretum secretorum et vlimum secretum.*" Bacon, *Liber sex scientiarum*, 185.

62. "*magnam prudentiam et sapientam . . . dei gratia adiutrice.*" Bacon, *Liber sex scientarium*, 185.

63. Matus, "Resurrected Bodies and Roger Bacon's Elixir," 329–31.

64. Burke, *Opus maius*, 585.

65. Ptolemy, *Liber quadripartiti Ptholemei; Centiloquium*, 107. Ptolemy, *Centiloquium*, MS Bayerische StaatsBibliothek, 2r.

66. Bonaventure, *Commentary on the Sentences*, Lib. IV, Dist. xlix, Part. ii, Art. i, Sect. i, Quaest. i, 582.

67. Bacon, *Liber sex scientiarum*, 185.

68. Bridges, *Opus maius*, II, 170–72.

69. Bacon, *Liber sex scientiarum*, 185–86.

70. Marrone, *In the Light of Thy Countenance*, II, 267–68.

71. This is the *De rerum proprietatibus* (*On the Properties of Things*) of Bartholomaeus Anglicus, an English Franciscan who died in 1272. He likely composed the encyclopedia around 1240. To the best of my knowledge, no scholar has yet connected *Pro conservanda* to this text. The essential source on Bartholomaeus's encyclopedia is Meyer, *Die Enzylopädie des Bartholomaeus Anglicus*.

72. "*ut dicit Constant.*," "*teste Aristot.*," and passim. Vitalis of Furno (?), *Pro conservanda*, 5. Many of these citations, however, come from work copied from *On the Properties*

of Things, and as such should not be read as necessarily influencing the author of *Pro conservanda.* See below.

73. Bartholomaeus's *On the Properties of Things* was constantly reworked and adapted by other authors, and in such form was ascribed to various authorities, including Albert the Great and Vincent of Beauvais. For a wide-ranging list of reworked editions of *On the Properties of Things*, see Meyer, *Die Enzyklopädie des Bartholomäus Anglicus,* 149–89.

74. It is difficult even to make assumptions about what region of Europe might have been home to its author, given the general uniformity of medical learning throughout the Latin West. Siraisi, *Medieval and Early Renaissance Medicine,* 55. At best, we can discern that the text was written for a non-Italian audience, given that certain dietary habits of Italians require explanation. Vitalis, *Pro conservanda,* 124.

75. Aristotle is referred to often, notably in a discussion of how a natural etiology may be posited for the existence of cynocephales and other monsters. Vitalis, *Pro conservanda,* 198. Aristotle is also mentioned in the discussion of "waters," but not in "planets," which is unsurprising given the recent denunciation of Aristotelian cosmology at Paris. Still, many such mentions of Aristotle derive from copying sections of *On the Properties of Things.*

76. Siraisi, *Medieval and Early Renaissance Medicine,* 55–56. Being university educated, however, does not suggest an actual difference in the content of remedies or treatments, Siraisi, *Medieval and Early Renaissance Medicine,* 32–33. What points to Montpellier is the discussion of a theriac-like substance. See below.

77. The author notes a number of Italian remedies, including those particularly useful for warding off drunkenness and hangovers. Vitalis, *Pro conservanda,* 124.

78. I consulted the 1519 edition of *De rerum proprietatibus,* but since editions and paginations vary I will cite Book and Chapter of the work.

79. "*Si quis multum doleat caput, uel sit Phreneticus aut aliquo in membro dolorem paciatur, crura eius et brachia ligentur, stringenturque in tantum ut ex ligaturae strictura laedatur, tunc capitis dolor diminuitur. quia unum membrum alteri compatiatur, et ad constrinctum membrum spiritus confluunt et humores, propter quod caput alleuiatur. Dolor enim uni membro superueniens, alterius membri dolorem mitigat.*" Vitalis, *Pro conservanda,* 128.

80. "*Malefici utuntur Vpupa, quae auis plurimum habitat in sepulchris, antris uel in fimo. estque immunda ualde, plumis a capite exeuntibus galeata. Cuius sanguine si quis se inunxerit, dormiendo se demones suffocantes uidebit. cuius cor malefactoribus ualet, nam suis in maleficiis eo utuntur. Vespertilio animal simile muri, auis, habens pennas continuas, uolat in crepusculo admodum celeri uolatu, in scissuris habitat parietum, frigidissimae naturae, oleum bibit in lapidibus. quare sanguis uespertilionis super palpebras linitus, pilos non finit crescere, quia sua frigiditate poros oppilat, quibus oppilatis, pili non recrescunt. uide in dict. rana. uulpis.*" Vitalis, *Pro conservanda,* 189–90. I think it no coincidence that the two animals hoopoes (*upupa*) and bats (*vespertilius*), are listed one after another in *Properties,* from which much of the text has been lifted verbatim. Bartholomaeus Anglicus, *De Rerum Proprietatibus,* XII, 37; XII, 38. Note that the bat is present in the entry because it is paired with hoopoes in the passage of Leviticus.

81. See Chapter 4.

82. Bartholomaeus's text on gold is somewhat lengthier than the entry in *Pro conservanda.* Bartholomaeus Anglicus, *De rerum proprietatibus,* XVI, 4.

83. "*Sciendum quod falsarii argentum et aurum ponunt in loco frigido et humido, quia humorem attrahunt, tuncque plus ponderant. Cum autem ab humorem fuerint desiccata, suum ad pondus redeunt. Item si auro liquefacto gallorum ossa uel caponum seu gallinarum apponantur seu misceantur, aurum consumunt. Unde ossa gallinacea uenenum auri sunt.*" Vitalis, *Pro conservanda,* 54.

84. The bull is found in *Extravagantes Communes,* Tom. II, col. 1295–96. The text of the bull is reprinted with a French translation in Halleux, *Les Textes alchimiques,* 124–26; cited henceforth as John XXII, *De crimine falsi.*

85. "*quod non est in rerum natura esse verum aurum vel argentum sophistica transmutatione confingant.*" John XXII, *De crimine falsi,* 1295–96.

86. "*et non alias alchimitum fornacis ignem vulgum ignorantem eludant.*" John XXII, *De crimine falsi,* 1296.

87. Just what kind of metals remains an issue. J. R. Partington appears to read *fidis metallis* as *foedis metallis,* and translates it is as "base metal." Partington, "Albertus Magnus on Alchemy," 58. This creates a problem, however, as the adjective *fidus,* meaning credulous or believing, seems appropriate for a later use of this term:"*ut fidis metallis cudant publicae monetae characteres fidis oculis.*" John XXII, *De crimine falsi,* 1296. Robert Halleux, on the other hand, retains *fidis* in both cases, translating the phrase *fidis metallis* as *métal honnête.* Halleux, *Les Textes alchimiques,* 125. I believe Halleux's translation to be the more accurate.

88. Oldrado believed alchemists imitated nature with their art. Migliorino, "Alchimia lecita e illecita nel Trecento," 25–26.

89. "They say that from one species of metal, such as tin, can be produced other types of metal, such as gold. And this is not unsuitable, for we also see that whenever something dead is produced from a living thing, as we see from worms from which silk is produced and many other things and from the plants from which woad is made. This is so much more the case from metals in which there is greater similarity and harmony (*dicunt quod ex una specie metalli, scilicet stanno, potest alia species metalli, scilicet aurum, produci. Nec hoc est inconveniens, nam et videmus quod quandoque ex re mortua producitur vita, ut videmus in vermibus ex quibus producitur sericus et aliis pluribus et ex herba producitur vitrum. Multomagis ex metallis in quibus est maior convenientia et similitudo*)." Oldrado da Ponte, *Consilium* 74, quoted in Migliorino, "Alchimia lecita e illecita," 26. See also Newman, *Promethean Ambitions,* 57–58, and Newman, "Technology and Alchemical Debate in the Late Middle Ages," 440–41.

90. Chiara Crisciani, *Il Papa e l'alchimia,* 45–50.

91. "*Et si clerici fuerint delinquentes: ipsi ultra praedictas poenas priventur beneficiis habitis, et prorsus reddantur inhabiles ad habenda.*" John XXII, *De crimine falsi,* 1236.

92. "*quamplures esse solo nomine christianos, qui . . . cum morte foedus ineunt, et pactum faciunt cum inferno.*" John XXII, *Super illius specula,* in *Bullarium Romanum* IV, 315–16. On the pact, see Boureau, *Satan the Heretic,* 68–92.

93. "*imagines, annulum vel speculum vel phiala, vel rem quamcumque aliam magice ad daemones inibi alligandos.*" John XXII, *Super illius specula*, 315–16.

94. *Prodiens* is in John XXII, *Extravagantes Iohannis XXII*, 221–24.

95. It is important to bear in mind, however, that while much of this bull deals with the binding of demons and use of magic equipment or books therein, it does not necessarily mean that alchemy would always remain separate from magic. By the fifteenth century, illicit alchemists were punished for magical offenses. Nummedal, *Alchemy and Authority in the Holy Roman Empire*, 3.

96. Vitalis, *Pro conservanda*, 54. Bartholomaeus Anglicus, *De rerum proprietatibus*, XVI, 4.

97. *Secundum Aristotolem . . . aurum sicut cetera metalla materialiter procreatur ex sulphure subtili rubeo et de viuo argento subtili et albo.*" Bartholomaeus Anglicus, *De rerum proprietatibus*, XVI, 4. Vitalis, *Pro conservanda*, 55. Cf. Bartholomaeus Anglicus, *De rerum proprietatibus*, XVI, 8.

98. Vitalis, *Pro conservanda*, 54. Bartholomaeus Anglicus, *De rerum proprietatibus*, XVI, 4.

99. Vitalis, *Pro conservanda*, 55.

100. The section on planetary governance of metals is in Bartholomaeus Anglicus, *De rerum proprietatibus*, VIII, 22. The two passages are as follows. "*Planetarum coniuncto cum ipsis signis diuersos habent effectus. nam planetae si coniuncti fuerint in signis igneis, siccitatem significat atque famem, excessum siccitatis et caloris. Si in signis aeris, multitudinem significat uentorum. At in signis frigidis, grauitatem significat frigoris, et caloris temperamentum. Planetarum effectus uariatur in bono, malo, secundum bonitatem et malitiam signorum in quibus eorum fit coniunctio. Quia si planetae fuerint beneuoli, bonum praetendunt. si uero mali, praetendunt malum.*" Vitalis, *Pro conservanda*, 230. Compare the following passage from *Properties*: "*Attenenda est planetarum coniunctio in ipsis signis: quia si plures fuerint in signis aquaticis in reuolutione anni signet multitudienm pluuiarum. Si in igneis siccitatem et famem et excessu caloris et siccitatis. si in aeris multitudinem uentorum si in signis frigidis grauitatem frigoris et temperamentum caloris. Et iterum capite viii dicitur. Quarum effectus signorum intenduntur in operationibus suis ex coninunctione planetarum siue in bono siue in malo quia si fuerint boni planete bonum praetendunt, si fuerint mali etiam contrarium.*" Bartholomaeus Anglicus, *De rerum proprietatibus*, VIII, 22.

101. This reflects the organizational priorities of the author of *Pro conservanda* rather than a lack of reliance on Anglicus. His section on waters draws from books and chapters unaffiliated with the term in Anglicus. For instance, the opening section on waters in *Pro conservanda* on the relative warmth of waters according to seasons borrows from *Properties'* sections on winter and summer. Vitalis, *Pro conservanda*, 1; Bartholomaeus Anglicus, *De rerum proprietatibus*, IX, 7 and IX, 5.

102. The entries are far too extensive to list the numerous parallels. As an example, however, *Pro conservanda* lifts a number of passages on the properties of special fountains throughout the world that also appears in *Properties*. Vitalis, *Pro conservanda*, 10, Bartholomaeus Anglicus, *De rerum proprietatibus*, XIII, 1.

103. Evidently "in certain waters just like that from wells, beans cannot be cooked, since that water is coarse and earthy, and beans by their nature are coarse, cold, dry, compact, and hard to digest. But they should be cooked in fountain water which is refined, and it penetrates the coarse substance of the bean (*In quibusdam etiam aquis uelut putealibus non possunt decoqui legumina, quoniam illa aqua grossa est et terrestris, et legumina de natura sua sunt grossa, frigida, sicca, compacta, et ad digerendum dura. Sed decoquantur in aqua fontana quae subtilis est, et grossam leguminis substantiam penetral*)." Vitalis, *Pro conservanda*, 11. Cf. "*Est autem legumen substantialiter grossum, frigidum, et siccum ad dirigendum durum et compactum; unde propter suam compactionem difficiles sunt ad coquendum nec etiam possunt bene decoquere in aqua puteali imo ad perfectam decoctionem indigent aqua fontana siue fluuiali.*" Bartholomaeus Anglicus, *De rerum proprietatibus*, XVII, 95.

104. Vitalis, *Pro conservanda*, 12.

105. "*si pilis capitis superfundatur, uel panno, et applicetur candela ardens, statim inflammabitur, uidebunturque pili capitis uel pannus ardere, quamuis de panno aut pilis nihil consumetur, flamma autem tam diu durabit, donec a flamma aqua illa consumpta sit.*" Vitalis, *Pro conservanda*, 12.

106. "*Recipe uinum bonum ac forte, purum, rubeum, et pone in alembico, et distilla lento igne. Sic fit aqua rosacea, et exibit per sublimationes aqua ardens, et si saepius distilletur, quanto plus distillabitur, tanto erit subtilior ac utilior.*" Vitalis, *Pro conservanda*, 12.

107. This is echoed in a later passage, where the author discusses how to seal the openings of an alembic with a paste made from wheat flour, "excepting that [aperture] which occupies the small channel which hangs down from the alembic (*excepto illo quod occupat illa parua canalis quae dependet ex alembico*)." Vitalis, *Pro conservanda*, 13.

108. "*Mercuriam congelat, cuprum dealbat. spiritus et corpora calcinata soluit.*" Vitalis, *Pro conservanda*, 12.

109. Wilson, *Philosophers, Iōsis and Water of Life*, 42–43.

110. Wilson, *Philosophers, Iōsis and Water of Life*, 46–47.

111. Thomson, "The Text of Michael Scot's Ars Alchemiae," 523–59, cited in Wilson, *Philosophers, Iōsis and Water of Life*, 92. Thomson, "Michael Scot's Ars Alchemiae," 555.

112. Wilson, *Philosophers, Iōsis and Water of Life*, 67.

113. "*omne metallum . . . soluit seu liquefacit,*" Vitalis, *Pro conservanda*, 13.

114. "*fertur quia habere xl uirtutes seu efficatias.*" Vitalis, *Pro conservanda*, 12.

115. "*narium faetorem, gingiuarum, ascellarumque tollit.*" Vitalis, *Pro conservanda*, 12.

116. "*apostema rumpit interius et exterius si potetur, si quia apostema exterius cum ipsa liniatur aliquoties. Oculorum maculam, eorumque ruborem et calorem aufert. . . . hominem super omnia reddit hilarem, . . . salsum phlegma, guttamque rosaceam curat, dentium dolorem remouet, . . . apostema in gutture rumpit, si frequenter gargarizetur.*" Vitalis, *Pro conservanda*, 12.

117. Vitalis, *Pro conservanda*, 13.

118. Wilson, *Philosophers, Iōsis and Water of Life*, 73–90.

119. Wilson, *Philosophers, Iōsis and Water of Life*, 73–90.

120. "*uulnera curat . . . Paralysim curat, ubi paralyticum membrum frequenter cum ea liniatur. Intellectum acuit si sobrie sumatur, praeterita ad memoriam reuocat. . . . iuuentutem*

conseruat, et senium retardat, . . . melancholicis plurimum confert, hydropicos frigida ex causa curat." Vitalis, *Pro conservanda*, 12.

121. Vitalis, *Pro conservanda*, 13.

122. For an introduction to the theriac in the Middle Ages, see Fabbri, "Treating Medieval Plague: The Wonderful Virtues of Theriac," 247–83.

123. Galen's text on theriac did not appear in Latin until the fourteenth century, but works on theriac by Avicenna and Averroës circulated in the late thirteenth. McVaugh, "Theriac at Montpellier," 113–16.

124. Fabbri, "Treating Medieval Plague: The Wonderful Virtues of Theriac," 254.

125. McVaugh, "Theriac at Montpellier," 113.

126. Vitalis, *Pro conservanda*, 13.

127. McVaugh, "Theriac at Montpellier," 114.

128. McVaugh, "Theriac at Montpellier," 125.

129. Fabbri, "Treating Medieval Plague: The Wonderful Virtues of Theriac," 252, 254.

130. "*Qui mirae autem et uirtutis et efficaciae aquam facere cupis. Recipe nucis muscatae, gariophyli, galangae, cardamomi, granorum paradisi, zinziberis, cinnamomi ana, ista in mortario pulverisentur, et ille pulueres in praedicta aqua ardenti ponantur, quae tandem aqua cum praedictis pulueribus sic commixta, lento cum igne distilletur in alembico.*" Vitalis, *Pro conservanda*, 13.

131. "*uirtutem herbarum omnium extrahit si in ea ponantur,*" Vitalis, *Pro conservanda*, 13.

132. Fabbri, "Treating Medieval Plague: The Wonderful Virtues of Theriac," 255. "*figitque spiritus non fixos,*" Vitalis, *Pro conservanda*, 13.

133. See, for instance, Multhauf, "John Rupescissa and the Origin of Medical Chemistry"; Calvet, "À la recherche de la médicine universelle. Questions sur l'élixir et la thériaque au 14e siècle"; Crisciani e Pereira, *L'Arte del Sole e della Luna, Alchimia e filosofia nel medioevo.*

134. Rupescissa, *De consideratione*, 51, 76, where John denigrates alchemical gold. There remains an ongoing problem of disentangling Rupescissa's alchemical work from that of Raymond Lull, or at least the Lullian school. Lynn Thorndike has pointed out, as have scholars before and since, that Rupescissa's work on the quintessence was circulating under the name of Raymond Lull as early as the end of the fourteenth century. The "adoption" of Rupescissa's work by the Lullian school no doubt was accompanied by interpolation and excision, which the manuscript tradition seems to bear out, leading George Sarton to declare the process practically hopeless. In assessing the tradition, Thorndike waded through much more of the material than did Sarton, and used not only the Basil printing of 1561, but also the shortest of manuscripts he had identified (a manuscript from the late fifteenth or early sixteenth century) and a longer, well-arranged copy from the fourteenth century. Thorndike's selections are made on aesthetic grounds—namely, he prefers manuscripts that are easier to read, more direct, and convey a "favorable impression," rather than texts with "pious cant" and "mystical smalltalk." While there is

no doubt that the Lullian school affected the manuscript tradition of Rupescissa's *De consideratione*, Thorndike's denigration of "pious cant" and "mystical smalltalk" does not necessarily argue against Rupescissa as author. Thorndike, *History of Magic and Experimental Science*, III, 355–58. We know from texts such as that of Bernard of Treves that alchemical texts were often used like recipe books (Bernard used John's in such a way), and the condensation of a work or elision of various elements viewed as unnecessary are just as likely as interpolation by later copyists. Bernard of Treves, *De chymico miraculo*. For discussion of this text, see Thorndike, *History of Magic and Experimental Science*, III, 628. That being said, having made some comparative notes with an early manuscript from the end of the fourteenth or early fifteenth century (Houghton MS Lat 220), I have decided, as have scholars such as DeVun and Michela Pereira, and even Thorndike at times, to cleave provisionally to the 1561 Basil edition, though it is relatively late. My decision stems from the fact that the Basil edition of 1561 adheres to a more reliable test of Rupescissa's authorship pointed out by Thorndike, namely that the text is concerned with the making the quintessence, an elixir of health, and not with transmutation of metals. Indeed the former is Rupescissa's premise for writing the work. Here I differ with more recent scholarship of Giancarlo Zanier. Zanier states that Rupescissa's aim is to make potable gold (or metals). This argument conflates the method with the aim, and ignores both the apocalyptic and the personal undercurrents of the work. Zanier, "Procedimenti farmacologici e pratiche chemioterpeutiche nel *De consideratione quintae essentiae*," 162. The implication then is that transmutation of metals is at least of one the Lullian interpolations of the text. In the 1561 edition, discussion of transmutation is hardly present, and in fact arises most conspicuously when Rupescissa argues that "*aurum alchimicum* (alchemical or transmuted gold)" is insufficient for work with the quintessence. *De consideratione*, 23.

135. "*non dico ut propheta, quia propheta non sum* . . . (Zech 13:5)" This declaration is found at the beginning of John's massive visionary work, *The Book Revealing the Times That Hasten to Appear* (*Liber ostensor quod adesse festinant tempora*), a massive tome composed in 1356 during the period of John's captivity in Avignon. The work is a treatise on coming events, mingling commentary on scripture and well-known prophecies with his personal assertions about the future. For an introduction see Rupescissa, *Liber ostensor*, 12–26, 45–62. John's declaration is found in Rupescissa, *Liber ostensor*, 108. He makes the same argument in Rupescissa, *Liber secretorum eventorum*, 213, 214. Cited hereafter as Rupescissa, LSE. Robert Lerner has argued that John "appears to be expressing himself as a prophet pure and simple." Robert Lerner, "Historical Introduction" in LSE, 40.

136. Rupescissa, *De consideratione*, 11.

137. For DeVun's discussion of this term, see DeVun, *Prophecy, Alchemy*, 106–7.

138. See both Rupescissa, *De consideratione,* 103–5, where John links himself to the book of secrets tradition, and DeVun, *Prophecy, Alchemy*, 3, 58–60 for his relationship to prior and later alchemists.

139. Rupescissa, *De consideratione*, 11. John's period of study likely would have been restricted once he took the Franciscan habit, for the various religious orders denied their

members access to the secular schools at the university. Smith, *The University of Toulouse in the Middle Ages*, 85.

140. Rupescissa, *De consideratione*, 13.

141. *De consideratione* (1351/2) contains allusions to his captivity, but does not specify which prison, exactly. *De consideratione*, 145. On the difficulties of sorting out John's sentence, see Lerner, "Historical Introduction," 30–31, 31n. Lerner notes that he has found only one contemporary witness to John's sentence, in the chronicle of Konrad of Halberstadt who was present in Avignon. Jean de Venette (d. 1368/9), the head of the Carmelite Order in France writes more vaguely about John, "Many questioned whether he were not mistaken or were not telling falsehoods or were not speaking by some pythonic or evil spirit. (*Et dubitabatur a multis ut fallet uel ut dicet mendicia uel [loqueretur] a quo spiritu pytonico seu maligno.*)" Generally, however, the chronicler offers a guarded endorsement of John's piety and prophetic powers. Venette, *The Chronicle of Jean de Venette*, 61, and Venette, *Cronica*, MS Arundel 28, 7v. Note that the final phrase beginning from "and speaking by some pythonic . . ." is found in the margins of the MS, placed there by the corrector and Newhall suggests that the "loqueretur" was omitted, though its omission would not alter the general meaning of the passage. Venette, *Chronicle of Jean Venette*, 211.

142. *De consideratione*, and to a lesser extent, the *Liber lucis*, have remained until fairly recently firmly in the province of the history of science. For most of the twentieth century John was more likely known among historians of chemistry and alchemy thanks to the massive surveys of the history of science from the early- and mid-twentieth century: Lynn Thorndike's *History of Magic and Experimental Science* and George Sarton's *Introduction to the History of Science,* completed in 1948. John's alchemical work maintains a significant position as part of broader fourteenth-century advances in medicine and pharmacology, in particular John's use of distillation to create medicines. DeVun, *Prophecy, Alchemy*, 5–6. Sarton, *Introduction to the History of Science*, III, 1572–74. Multhauf, "John of Rupescissa and the Origins of Medical Chemistry," 359–67. Sarton's and Thorndike's studies have their flaws, in particular a teleological orientation in terms of the development of science that often leads to the dismissal of some texts and authors, as well as an inability to explain fully contradictions in scientific development. An example of this is Sarton's interpretation of two bulls issued early in the Avignon papacy. The first, *Deus scientiarum*, was issued by Clement V in 1309. Sarton argues that the bull is evidence of a broad reorganization of medical learning on Arabo-Greek grounds based on newly translated texts (*Introduction to the History of Science*, III, 44). The bull in fact grants to the faculty of Montpellier the right to certify practitioners of medicine. "*Nos igitur intendentes quod in electione cancellarii supradicti, qui, sicut asseritur, studio presidet memorato, et certam potestatem seu iurisdictionem habet in illo.*" Clément V, "*Deus scientiarum*," cap. 1065, f. 236b. The second bull is *De crimine falsi*, issued by John XXII in 1317. Sarton read these bulls as evidence for steady progress in the professionalization of the medical field. Sarton, *Introduction to the History of Science*, III, 44. Clement's bull fits with this thesis, but has as much to do with the assertion of university privileges as with professionalization. John's bull was aimed at forgers of currency rather than alchemists per se, as discussed earlier. Only a notion of alchemy as accreted superstition or pseudoscience justifies

a conclusion that its condemnation professionalized scientific practice. Thankfully, more recent scholarship, such as that undertaken by Michela Pereira, has gone further in establishing the significance of *De consideratione* as a groundbreaking treatise. "Only in the work of John of Rupescissa [were] the medical tradition and the alchemical tradition of distillation conclusively consolidated. (*La tradizione medica e quella alchemica della distillazione si unificheranno definitivamente solo nell'opera di Giovanni da Rupescissa.*)" Pereira, "L'Alchimista come medico perfetto," 93. Yet it was not until Leah DeVun's discussion of its apocalyptic resonances that *De consideratione* began to be viewed as something more than a link in the chain between alchemy and what Robert Multhauf termed "medical chemistry." DeVun, *Prophecy, Alchemy*, 6. Multhauf, "John of Rupescissa and the Origins of Medical Chemistry," 359.

143. John has both of these meanings in mind. Rupescissa, *De consideratione*, 19.

144. Rupescissa, *De consideratione*, 116–18.

145. Rupescissa, *De consideratione*, 20. John uses the term quintessence almost interchangeably with *aqua ardens*, though he points out that the latter is a manifestation of the former, and also is synonymous with other, more typical alchemical products, including *aqua vitae* (water of life) and *lac virginis* (milk of the virgin). This point is taken up in the following section.

146. Rupescissa, *De consideratione*, 22–29.

147. Some discussion of the particulars of John's distillatory method is found in Zanier, "Procedimenti farmacologici," 165–73.

148. Rupescissa, *De consideratione*, 20.

149. See Wilson, *Philosophers, Iōsis and Water of Life*, as well as standard texts such as Holmyard, *Alchemy*, Berthelot, *La chimie au moyen age*, and Ganzenmüller, *Die Alchemie im Mittelalter*. More specific texts on the *aqua ardens* include Diels, *Die Entdeckung des Alkohols* (Berlin, 1913); Gwei-Djen, Needham, and Needham, "The Coming of Ardent Water," and Forbes, *Short History*.

150. Thorp, "The Luminousness of the Quintessence," 105. Regarding Aristotle's innovative understanding of the stars, see also Elders, *Aristotle's Cosmology* 10–16, 34–38, 88, 214–17.

151. Plato, *Timaeus*, 40a–b. I have used Zeyl's translation, but textual citations are standard.

152. Grant, *Planets, Stars, and Orbs*, 189, 192.

153. Aristotle, *De caelo*, 269b. Aristotle's proof of a heavenly substance, in fact, depends on its removal to the heavens. Aristotle conceived of the universe as spherical (and, hence, spatially finite), with the earth at the center and the heavens at the margins. Thus, the perfect, immutable substance of the heavens lies at the farthest remove from the earth, which shares none of these qualities: "the superior glory of its nature is proportionate to its distance from this world of ours." See also Leggatt, "Introduction" in Aristotle, *On the Heavens I & II*, 11. An English translation can be found in Leggatt's volume, henceforth cited as Aristotle, *On the Heavens* when explicitly used. Otherwise, general references will be to Aristotle, *De caelo*.

154. Luminescence is produced, instead, by friction of celestial bodies with air, rather than any innate fiery composition of the heavenly bodies themselves. It is the air itself, which combusts from this friction. Aristotle, *De caelo*, 289a.

155. Thomas's objection stems not so much from Aristotle's argument on the distance of heaven from earth, but on another aspect of the Philosopher's proof that heaven is divine, namely that heaven is spherical and moves in a circle (circularity being primary and perfect). (See Aristotle, *De caelo*, 269b, 286b). "For [Aristotle] proceeds from the perfection of the circular line to proving the perfection of a circular motion and from its perfection he proceeds to proving the perfection of a circular body. And thus it seems his proof is circular, since a circular line does not seem to be anything other than that of an actual body which is moved circularly. (*Procedit enim ex perfectione lineae circularis ad probandum perfectionem circularis motus; ex cuius perfectione procedit ad probandum perfectionem circularis corporis; et sic videtur esse alia quam quae est ipsius corporis quod circulariter movetur.*)" Aquinas, *In Aristotelis libros De caelo*, L.I, 1.iv [45].

156. Aquinas, *In Aristotelis libros De caelo*, L.I., 1.iv. [47, 49] Thomas is dealing again with circular motion. He takes issue with Aristotle's argument that something outside of nature is against nature ("*Philosophus . . . accepit praeter naturam pro eo quod est contra naturam*"). Thomas asserts that the motion of heaven is according to nature, "since it is irrational that that which is everlasting be outside nature and a movement that is not everlasting be according to nature (*irrationabile autem est quod id quod est sempiternum, sit praeter naturam, et motus non sempiternus sit secundum naturam*)."

157. Aristotle, *On the Heavens*, 289a. "It is most reasonable, then, and a consequence of what we have said, to make each star from that body in which it happens to be moving, since we said that there is something which is of nature to move in a circle. For just as those who declare the stars are fiery do so because they maintain that the upper body is fire and that it is reasonable that they are formed from the elements in which each one is located, so do we make a similar claim."

158. "*non sint simplicia corpora.*" Aquinas, *In Aristotelis libros De caelo*, L.II, 1.x [384–85].

159. This was the Platonic conception (found in *Timaeus*) that predominated in the Middle Ages prior to the reintroduction of Aristotle. Grant, *Planets, Stars, and Orbs*, 189–90.

160. "*non omnis diversitas, proprie loquendo, habet rationem contrarietatis.*" Aquinas, *In Aristotelis libros De caelo*, L.II, 1.x [384].

161. "*in corporibus caelestibus, quanto est maior congregatio per modum inspissationis, tanto magis mutliplicatur luminositas et virtus activa, sicut patet in ipsis corporibus.*" Aquinas, *In Aristotelis libros De caelo*, L.II, 1.x [384].

162. "*lumen non est proprium ignis, sed est aliquid commune sibi et supremo corpori.*" Aquinas, *In Aristotelis libros De caelo*, L.II, 1.x [390].

163. "*Aristoteles . . . ostendit ex motu stellarum igniri inferiora corpora; in igne autem invenietur calor et lumen.*" Aquinas, *In Aristotelis libros De caelo*, L.II, 1.x [390].

164. To be sure, Thomas differs with Aristotle on other important ideas, for instance the Philosopher's claim that the heavens were ungenerated. While this is an important

distinction in terms of Christian dogma, it, like most other points of contention, has little bearing on John's alchemical endeavors. Grant, *Planets, Stars, and Orbs*, 192.

165. Aquinas, *In Aristotelis libros De caelo*, L.I, 1.vi [63].

166. Peter of Auvergne, *Questions on Aristotle's* de Caelo, 176–77. Opponents of the idea that heaven consisted of matter included Jean Buridan, the Franciscan Peter Aureol, Albert of Saxony, and Godfrey of Fontaine. See the discussion in Grant, *Planets, Stars, and Orbs*, 245–49.

167. Grant, *Planets, Stars, and Orbs*, 254–59.

168. *De generatione stellarum*, in (pseudo-) Robert Grosseteste, *Die philosophischen Werke des Robert Grosseteste*, 36, quoted from Pereira, "Heavens on Earth," 139–40.

169. "The Most High imbued the fifth essence not only in fiery water, but indeed in everything . . . (*Concreauit Altissimus non solum in aqua ardente quintam Essentiam, sed etiam in omni* . . .)" Rupescissa, *De consideratione*, 36.

170. Pereira, "Heavens on Earth," 140.

171. John himself believes that that he is revealing something new, telling his readers they can hide their knowledge in plain sight: "And when you wish to hide it, call it the quintessence, since no great Philosopher wants to spread its nature and name about, but rather to bury the truth among themselves (*Et quando tu voles eam occultare, vocabis eam quintam Essentiam, quia hanc eius naturam et hoc nomen suum summi Philosophi nemini pandere voluerunt sed secum veritatem sepeliri fecerunt*)." Rupescissa, *De consideratione*, 19.

172. Rupescissa, *De consideratione*, 20.

173. For the still, see Rupescissa, *De consideratione*, 31. "*fabricari facies in furno vitreariorum vnum distillatorium tale: totum integrum, ex vna pecia cum vno solo foramine superius, per quod ponetur & extrahetur aqua: Vides enim sic mirifice instrumentum formatum, vt illud quod virtute ignis ascendit & distillatur in vas per canales brachiales, iterum reportetur, vt iterum ascendat, & iterum descendat continue die ac nocte, donec aqua ardens in quintam Essentiam per Dei voluntatem caelitus conuertatur.*" On the process of distillation see Rupescissa, *De consideratione*, 29–30. "*Non accipies vinum nimis aquosum, nec vinum nigrum, terrestre, insipidum: sed vinum nobile, iocundum, saporosum, & odoriferum, melius quam inueniri possit: & distilla eum in canonibus totiens, donec feceris meliorem aquam ardentem quam scias facere, vt puta a tribus vicibus vsque septies eam distilles: & haec est aqua ardens.*"

174. "*Philosophi autem vocaverunt caelum quintam Essentiam respectu quatuor elementorum, quia in se caelum, est incorruptibile et immutabile.*" Rupescissa, *De consideratione*, 19.

175. "*Et quod non sit humida sicut elementum aquae, demonstratum est, quia comburitur, quod aquae elementari repugnat. Quod non sit calida et humida sicut aer, declaratur, quia aer siccus quacunque re corrumpitur, sicut patet in generatione aranearum et muscarum: et ideo ista semper incorrupta manet, si clausa a volatu servetur. Quod non sit sicca et frigida sicut terra, demonstratur expresse: quia summe acuit, summe & calefacit. Quod autem non sit calida et sicca sicut ignis, demonstratur ad oculum: quia calida infrigidat, et calidos morbos minuit et annullat, sicut infra probabo.*" Rupescissa, *De considerationee*, 20–21.

176. "*creavit Altissimus ad conservationem quatuor qualitatum corporis humani, sicut caelum ad conservationem totius universi,*" Rupescissa, *De consideratione*, 21. Heaven's role in preserving the universe was a widespread scholastic notion in the fourteenth century. Grant, *Planets, Stars, and Orbs*, 615–17.

177. Rupescissa, *De consideratione*, 18–19.

178. "*universae stellae caeli habent suam influentiam singularem ex iussu et ordinatione Dei, et quaelibet stella suam proprietatem habet super rem determinatam et certam: ut stella poli super Adamantem et ferrum: et ita ius Lunae super aquam maris, Sol super aurum: Luna super argentum: imagines hominum caeli super corpora humana: imago arietis caelo* super arietes terrestres et caput hominis." Rupescissa, *De consideratione*, 25–26.

179. Calvet, "À la recherche de la médicine universelle," 182–83. Though John rarely discusses his sources in *De consideratione*, he does reference "Master Arnold (*Magister Arnaldus*)" and the *Rosarius* (an alchemical work bearing Arnold's name, but of only doubtful provenance) in the *Liber lucis*. Rupescissa, *Il libro della luce*, 137.

180. Rupescissa, *De consideratione*, 28.

181. "*et omnem excessum cuiuscunque qualitatis amputet, et quamcunque qualitatem deperditam restauret.*" Rupescissa, *De consideratione*, 18.

182. Rupescissa, *De consideratione*, 16–18, esp. "*Restat ergo rem quaerere, quae citra terminum vitae nostrae a Deo praefixum, possit corpus nostrum sine corruptione servare, sanare, et conservare, infirmum curare deperditum restaurare, donec veniat ultima dies mortis in termino praefixo a Deo.*" John also diligently quotes Hebrews 9:27 ("It is appointed for men to die") and Job 14:5 ("Short are the days of man"). In discussing one's preordained end, John takes up the hackneyed, if amusing, argument that if nothing else kills you, there's always a bolt of lightning. Rupescissa, *De consideratione*, 16–18.

CHAPTER 3. THE APOCALYPTIC IMPERATIVE

1. See especially Burr, *The Spiritual Franciscans* and Burr, *Olivi's Peacable Kingdom*, as well as Lambert, *Franciscan Poverty*.

2. DeVun, *Alchemy, Prophecy*, 102.

3. Bynum, *Christian Materiality*, 18, 219.

4. See "Admonitions" in Francis of Assisi, *The Writings of Saint Francis*, 8. Francis expected that his brothers would "preach by their works" and never strive to displace the secular clergy. Indeed, if a brother was prohibited from preaching, he was to meekly and obediently accept the charge. See "Rules" in Francis of Assisi, *The Writings of Saint Francis* 50–52, 71.

5. "And I strictly enjoin on all my brothers, clerics and laics, by obedience, not to put glosses on the Rule or these words [of the Testament] saying: Thus they ought to be understood." "Testament of the Holy Father Saint Francis" in Francis of Assisi, *The Writings of Saint Francis*, 86.

6. The bull is in *Bullarii Franciscani Epitome*, IV, 229a. See also Burr, *The Spiritual Franciscans*, 2–4, 12–14.

7. "Brothers . . . that you may know what kind of life the present and future brothers are to live, understand the truth of things that are to come. We will find now, at the beginning of our life, fruits that are extremely sweet and pleasant to eat; but a little later some that are less sweet and less pleasant will be offered; and, lastly some that are full of bitterness will be given, which we will not be able to eat, for because of their bitterness they will be inedible to all, though they will manifest some external fragrance and beauty." I Celano Bk 2, cap XI, par 28 in Thomas of Celano, *St. Francis of Assisi: First and Second Life of St. Francis*.

8. II Celano bears the mark of the *Legenda Trium Sociorum* (*Legend of the Three Companions*), also called *Scripta Leonis, Rufini, et Angeli* (*Writings of Leo, Rufinus, and Angelo*). These remembrances of Francis's companions are stoked with imagery of protean Franciscan life, idealized in its simplicity. See Moorman, *The Sources for the Life of St. Francis of Assisi*, and Brooke, *Scripta Leonis, Rufini et Angeli*, 2–17, 20–66. It is important to bear in mind, however, that at this point no reified factions existed within the Order. Burr, *The Spiritual Franciscans*, 12–37.

9. There are a number of exceptional studies on Joachim of Fiore and Joachism. See Reeves, *Joachim of Fiore*; Reeves, *The Influence of Prophecy*; McGinn, *The Calabrian Abbot*.

10. McGinn, *Antichrist*, 153.

11. Reeves, *The Influence of Prophecy*, 145–49, 242–50.

12. Bale, a former Carmelite converted to the beliefs of Martin Luther and Philipp Melancthon, used Joachite exegesis of Revelation to reinterpret the history of English heretics, such as John Wyclif and the Lollard John Oldcastle, as forerunners of Protestantism. Bale's Joachite works include *The Actes of English Votaries or Unchast examples of the Englsih Votaryes*, "The Image of Both Churches," and "The Examination and Death of Lord Cobham." For discussion of Bale, see Firth, *The Apocalyptic Tradition in Reformation Britain*; Fairfield, *John Bale, Mythmaker for the English Reformation*.

13. Rupert of Deutz (1075–1129), writing a generation before Joachim, also posited a three- and sevenfold view of salvation history based on his reading of the Apocalypse, though he made no prophecy of a third *status*, and the tenor of Rupert's work was "much more ecclesiological than eschatological." See Van Engen, *Rupert of Deutz*, 275–82. Hildegard of Bingen (1098–1179), a contemporary of Joachim's, also conceived of history as trinitarian in structure, and, like Rupert, used her apocalyptic visions to comment on current ecclesiological and political issues. Unlike Joachim, Hildegard did not hold a notion of a third *status*. See Kienzle, *Hildegard of Bingen and Her Gospel Homilies*, 157, 163–97. For a discussion of the uniqueness of Joachim's conception of a new *status*, see below.

14. For instance, Joachim characterizes the first *status* as the era when "the order of the married shone forth" (*claruit ordo coniugatorum*), "the order of clerics" (*ordo clericorum*) represents the second, and "the order of monks" (*ordo monachorum*) the third. Joachim of Fiore, *Liber de concordia*, 405–6.

15. Joachim of Fiore, *Liber de concordia*, 405–6.

16. Some discussion of the scheme presented above can be found in Reeves, *Joachim of Fiore*, 29–55.

17. Reeves, *Joachim of Fiore*, 7.

18. For a discussion of Joachim's patterns, see Reeves, *Joachim of Fiore*, 8–9.

19. "It is known that there are many antichrists [*Constet . . . multos esse antichristos*]." Joachim of Fiore, *Expositio in Apocalypsim* (Venice, 1527; repr. Frankfurt: Minerva, 1964), f.10r, col i; McGinn, *Antichrist*, 136; McGinn, *The Calabrian Abbot*, 110–11; Lerner, "Antichrists and Antichrist," 559–63.

20. "*septimus rex qui proprie dicitur Antichristus*." See the representation of the *figura* in McGinn, *The Calabrian Abbot*, 111.

21. Joachim of Fiore, *Expositio in Apocalypsim*, f. 10r, col. i–ii.

22. "*Gog ultimus Antichristus*," quoted in McGinn, *The Calabrian Abbot*, 111. This term from the *figurae* is echoed in Joachim of Fiore, *Expositio in Apocalypsim*, f. 10r, col. i.

23. Joachim of Fiore, *Expositio in Apocalypsim*, f, 11v, col. i. McGinn, *Antichrist*, 139–41.

24. Elijah was carried to heaven by a fiery chariot (2 Kings 2:11). Enoch's fate was similar, but put in more vague terms, namely that he did not die, but "God took him away" (*tulit eum Deus*) (Gen 5: 24), a statement echoed in the New Testament (Hebrews 11:5).

25. Joachim does use Enoch and Elijah to describe the spiritual men. Joachim of Fiore, *Expositio in Apocalypsim*, ff. 137r–139v, 146v–150v, esp, f. 149r, col. ii. (The passages glossed by Joachim are Rev 10: 1–3 and Rev 11: 3–6.)

26. No single factor explains why the Dominican Order resisted the Joachite prophecy in a way many Franciscans did not. One can point, however, to Thomas Aquinas's studied opposition to Joachism as a significant factor, in addition to the more general rivalry with the Friars Minor.

27. McGinn, *The Calabrian Abbot*, 113.

28. McGinn, "Pastor Angelicus," 224–25.

29. "*ascendet . . . uniuersalis scilicet pontifex noue Ierusalem, hoc est sancte matris ecclesie . . . dabitur ei plena libertas ad innouandam christianam religionem et ad predicandum uerbum dei*." Joachim of Fiore, *Liber de concordia*, 402.

30. Joachim of Fiore, *Liber de concordia*, 402.

31. McGinn, "Pastor Angelicus," 225–26; Joachim of Fiore, *Liber de concordia*, 402.

32. McGinn, *The Calabrian Abbot*, 112–13.

33. "*maximi Joachite*." The description comes from the chronicle of Salimbene de Adam, a former adherent of Joachism. Salimbene uses the Latin plural since he is discussing not only John, but also his friend Hugh of Digne. Salimbene, *Cronica*, 334.

34. Anderson, *A Call to Piety*, xi–xiii.

35. The hesitancy to discuss Joachim in the wake of the scandal of the Eternal Gospel is found throughout the Franciscan Order during this period. Whalen, *Dominion of God*, 192.

36. Bacon mentions Joachim in his own work, notably in his *Opus maius* and the *Compendium studii philosophiae*. I discuss both examples below. Davide Bigalli outlines Joachim's influence among English friars, including Adam Marsh and Robert Grosseteste, in Bigalli, *I Tartari e L'Apocalisse*, 105–15.

37. For the papal role in converting the world in the Joachite structure, see McGinn, *The Calabrian Abbot*, 112. See below for a discussion of conversion. Whalen, *Dominion of God*, 190. In regard to this contribution, there is general agreement as to the influence Bacon had on subsequent apocalyptic development. Matus, "Reconsidering Roger Bacon's Apocalypticism," 190n, 197.

38. Burke, *Opus maius*, 407.

39. The epistle is in Bacon, *Opera quaedam hactenus inedita*, 1.

40. The letter is in Gasquet, "An Unpublished Fragment of a Work by Roger Bacon," 494–517. A reference to Antichrist comes early in the letter, on p. 498. Some additional discussion of the fragment comes from Easton, *Roger Bacon and His Search for a Universal Science*, 145–47. An identical passage is in the *Opus tertium*, though slightly later in the preamble. Brewer, *Opus tertium*, 8.

41. Burke, *Opus maius*, 415.

42. Burke, *Opus maius*, 290.

43. Emmerson, *Antichrist in the Middle Ages*, 79–107. McGinn, *Antichrist*, 86; McGinn, "Pastor Angelicus," 227. I have taken up this issue in more detail in Matus, "Reconsidering Roger Bacon's Apocalypticism," 198–206.

44. Joachim of Fiore, *Liber de concordia*, 402; McGinn, "Pastor Angelicus," 225; McGinn, *The Calabrian Abbot*, 112.

45. Roger Bacon, *Compendium studii philosophiae*, 227. Bacon's conversion program is rather untraditional, in that he purports to "make a plea for faith . . . through science, not by arguments, but by works, which is a far more effective way." There are two aspects to this. One is to awe an unbeliever with the powers of science, so that a "subdued" mind "may believe [religious truths] although it does not understand them." Second, Bacon advocates the use of science "to separate the illusions of magic and to detect all their errors," by which they can strip away the trappings of false faith just as much as they demonstrate the authenticity of Christian miracles. Burke, *Opus maius*, 632–33.

46. Daniel, *The Franciscan Concept of Mission*, 64–65. Bartlett, *The Natural and Supernatural*, 102–3.

47. Little, *Opus tertium*, 9–12.

48. Here I feel the need to emphasize Bacon's use of the Latin comparative *magis* (more), which implies that everyone outside Latin Christendom should be assumed to be at least a little hostile. Little, *Opus tertium*, 11.

49. "*propter violentiam gentium que invaderunt mundum, ut sunt Judei inclusi in montibus Hircanorum et Gog et Magog, et naciones incluse ab Alexandro ad portas Caspias, et propter Antichristum et suos.*" Little, *Opus tertium*, 11.

50. Gow, *The Red Jews*, 24–25.

51. Bridges, *Opus maius*, I, 365.

52. The Jews of the Hyrcanian mountains is a reference to the idea that the lost tribes of Israel, who by at least the era of Peter Comestor (d. 1178/9) were conflated with armies of Antichrist. Gow, *The Red Jews*, 38.

53. Bigalli, *I Tartari e l'apocalisse*, 7–33. Little, *Opus tertium*, 13. At the same time, Bacon has a rather unrealistic view of their military prowess: "the Tatars are small, weak men, and hardly eat or drink anything to strengthen their nature, [they are] not swift of foot, and properly speaking unarmed except for arrows with which to frighten those whom they pursue. . . . Hence their success must be due to the wonderful works of science by means of which they tread the world under foot." Burke, *Opus maius*, 416. This generally reflects Bacon's opinions elsewhere regarding Alexander's conquests, for which Aristotle's science receives the lion's share of the credit, rather than the military strength of Alexander's army. Burke, *Opus maius*, 408. Little, *Opus tertium*, 53.

54. Burke, *Opus maius*, 381–84.

55. Joachim considered Gog to be an Antichrist, but Bacon does not seem to do so. Joachim of Fiore, *Expositio in Apocalypsim*, f. 10, col. i.

56. "*Si ergo sciverimus ex qua parte isti venient, possumus considerare quod parte contraria veniet Antichristus.*" Little, *Opus tertium*, 11–12.

57. The novelty of this idea is evident in the famous anecdote from Roger of Hoveden regarding Joachim's meeting with Richard Coeur de Leon at Messina. After hearing Joachim's view, Richard expresses shock that Joachim differs from the established Antichrist legend: "I thought Antichrist was to be born in Babylon, or Antioch, from the line of Dan (*Putabam quod Antichristus nasceretur in Babylonia, vel in Antiocha, de stirpe Dan*)." "Benedict of Peterborough" (Roger of Hoveden), *The Chronicle of the Reigns of Henry II and Richard I*, II, 154. The same passage (substituting only *progenie* for *stirpe*) can be found in the later, expanded version of the chronicle: Roger Hoveden, *Chronica*, III, 78. On issues of authenticity and authorship of this account, see Reeves, *Influence of Prophecy*, 6–9, esp. 8; and Lerner, "Antichrists and Antichrist in Joachim of Fiore," 553n, 567n.

58. Burke, *Opus maius*, 208. Woodward and Howe, "Roger Bacon on Geography," 202–4.

59. Neff, "Lesser Brothers," 676.

60. The author and precise dating of the fresco cycle in the Upper Church of San Francesco remains a subject of (sometimes virulent) debate. An introduction to the historiography of this issue can be found in de Wesselow, "The Date of the St. Francis Cycle," 113–33.

61. Thompson, *Francis of Assisi*, 74–75.

62. Scrutiny began, perhaps not surprisingly, with the first Franciscan pope, Nicholas IV, who in 1290 took official action against rigorist members of his Order. This likely set a precedent for future popes to involve themselves in what had been an intramural dispute. Burr, *The Spiritual Franciscans*, 108. On the dating, see de Wesselow, "The Date of the St. Francis Cycle," 134, 167.

63. De Wesslow, "The Date of the St. Francis Cycle," 166.

64. McGinn, "Pastor Angelicus," 228.

65. Whalen, *Dominion of God*, 193.

66. "*praeveniet unus beatissimus papa, qui omnes corruptiones tollet de studio, et ecclesia, et caeteris, et renovetur mundus, et intret plenitudo gentium . . . ad fidem.*" Roger Bacon, *Compendium studii philosophiae*, 402. A. G. Little notes that Brewer's transcription in the *Compendium studii* appears to read Festo (here, *Festonis*) for Sesto. *Franciscan Papers*, 91. Little makes no definitive determination, but given Bacon's use of Pliny it seems clear that "Seston" from *Natural History* is likely the reference here. Certainly, it was read that way by Bridges and Burke in a parallel passage in the *Opus maius*. Burke, *Opus maius*, 290. Seston was accompanied by an eagle, which may mean that a more literal translation of *Aquila* is required here rather than its use as a proper name.

67. "*Hic Generalis frater Hieronymous de multorum fratrum consilio condemnavit et reprobavit doctrinam Fratris Rogerii Bachonis Anglici, sacrae theologiae magistri, continens aliquas novitates suspectas, propter quas fuit idem Rogerius carceri condemnatus, praecipiendo omnibus fratribus ut nullus illam teneret, sed ipsam vitaret, ut per Ordinem reprobatum.*" *Chronica XXIV Generalium Ordinis Minorum*, 360. See the discussion in Power, "A Mirror for Every Age," 659–60; and, Ingham, *The Medieval New*, 61–62.

68. "*operationes alchimiae, necromantiae, sortilegiorum vel superstitionum quarumlibet seu.*" Bihl, "Statuta Glia Ordinis 1260, 1279 et 1292," 35. Cf. de Dmitrewski, "Fr. Bernard Délicieux, O.F.M.: Sa lutte contre l'inquisition," 21.

69. Hackett, "Roger Bacon: His Life, Career, and Works," 9.

70. For the afterlife of Bacon's ideas, see Paravicini Bagliani, *The Pope's Body*, 204–11.

71. Burke, *Opus maius*, 415.

72. Burke, *Opus maius*, 634.

73. This is not to say that the rigorists thought well of him, either. Burr, *The Spiritual Franciscans*, 32–36.

74. During the 1270s and 1280s, there was no specific "Spiritual" party. Nor was Joachite speculation a common feature of all the rigorists, or at least not nearly as much as fidelity to the Brother Leo tradition, that is, the *Legend of the Three Companions* and other sources attributed to Leo and Francis's inner circle. See Burr, *The Spiritual Franciscans*, 49.

75. Burr, *The Spiritual Franciscans*, 51. Olivi's censure is thoroughly covered by Burr, *The Persecution of Peter Olivi*.

76. Burr. *The Spiritual Franciscans*, 51–53.

77. Burr, *Olivi's Peaceable Kingdom*, 66, 68–71, 75–77, 78.

78. Burr, *Olivi's Peaceable Kingdom*, 92.

79. Burr, *Olivi's Peaceable Kingdom*, 95.

80. Burr, *The Spiritual Franciscans*, 88.

81. Vincenzo Mauro, "La disputata 'de anima' tra Vitale du Four e Pietro di Giovanni Olivi," 89–138.

82. Burr, *Olivi's Peaceable Kingdom*, 44–54, quoted at 45. The commentaries in question are published in Alexander Halensis, *Commentarii in Apocalypsim* is found in Alexander Halensis, *Opera omnia,* and *Vidit Iacob* is in two *Opera omnia* of Aquinas. The Parma printing is found in Aquinas, *Opera omnia* 23, 325–511, and the Paris edition is in Aquinas, *Opera omnia* 31, 469–661.

83. Joachite material appears specifically in chapters on Revelation 13 and 17. Burr, "Antichrist and Islam in Medieval Franciscan Exegesis," 135. Burr, *Olivi's Peaceable Kingdom*, 45.

84. Burr, *Olivi's Peaceable Kingdom*, 46, 87–88.

85. "*In illa die tribulationum . . . Tunc aderit stultus plato cum discipulis. Tunc nec proderint aristotilli argumenta sua. Tunc adherit filius hominis,*" MS Assisi 66, 60v.

86. "*magnus heresiarcha, nocte lucens, in oscuris docens,*" MS Assisi 66, 74r.

87. A full bibliography of sources pertaining to the condemnations exceeds the confines of this project. As a beginning, see van Steenberghen, *Aristotle in the West*; Wippel, "The Parisian Condemnations of 1270 and 1277"; Wippel "The Condemnations of 1270 and 1277 at Paris"; Hissette, *Enquête sur les 219 articles condamnés à Paris*; Bianchi, *Censure et liberté intellectuelle à l'université de Paris*; Thijssen, *Censure and Heresy at the University of Paris.*

88. There are additional interpretations of the controversy. Takashi Shogimen sees the condemnations as the spark that led academic theologians to assert their doctrinal authority. Shogimen, "Academic Controversies," 233–47, esp. 233–35.

89. Thijssen, *Censure and Heresy*, 50; Wippel, "The Condemnations of 1270 and 1277 at Paris," 169–70; van Steenberghen, *Aristotle in the West*, 59–126, 230–38. Wippel, "The Parisian Condemnations of 1270 and 1277," 65–68.

90. It seems that no objection was leveled at Aristotle's ethics. Van Steenberghen limits the initial concerns over Aristotelianism to the eternity of the world and employing dreams to foretell the future. van Steenberghen, *Aristotle in the West*, 69.

91. Other followers of Aristotle, such as Albertus Magnus, also condemned David of Dinant. Van Steenberghen, *Aristotle in the West*, 69. The precise thought of both is unknown, as their books were burned and only hostile witnesses remain, but it is presumed that Amalric championed a kind of pantheism and that at some point David claimed God to be made of prime matter. The confusion among medieval observers is sorted out in Maccagnolo, "David Dinant and the Beginnings of Aristotelianism in the West," 429–36.

92. Oxford, where Bacon studied first, had no such prohibitions against the teaching of Aristotle, nor did the University of Toulouse, at least until 1245.

93. Lafleur and Carrier, *L'enseignement de la philosophie au XIIe siècle*, xi–xiv. Wippel, "The Parisian Condemnations," 67.

94. Wippel uses the term "Radical Aristotelianism" to refer to Siger of Brabant, Boethius of Dacia, and their followers. Wippel, "Condemnations of 1270 and 1277," 174.

95. Thijssen, *Censure and Heresy*, 50. Thijssen hypothesizes that neither Siger nor Boethius was named because they may have already been acquitted in a trial for heresy. If so, then naming them in the condemnations would have amounted to double jeopardy. Thijssen, *Censure and Heresy*, 47–48.

96. Wippel, "Condemnations of 1270 and 1277," 174.

97. Wippel, "Condemnations of 1270 and 1277," 174. Henry of Ghent was among the doctors consulted by Tempier prior to the condemnations. Thijssen, *Censure and Heresy*, 19. In an irony all too common in these events, Henry later was summoned to appear before Tempier and coerced into retracting some of his Thomist positions just weeks after the bishop issued the condemnations. Bianchi, "1277: A Turning Point in Medieval Philosophy?" 92. Marrone, *In the Light of Thy Countenance*, 267–68.

98. Wippel, "Condemnations of 1270 and 1277," 175. Bishop Tempier's letter underscores this: "For they [the philosophers] say these things are true according to philosophy but not according to the Catholic faith, as if there were two contradictory truths of scripture." Tempier, "Condemnations of 219 Propositions," 337.

99. van Steenburghen, *Aristotle in the West*, 223–24. An important corollary to this was the notion that the universe was eternal, which baldly conflicted with Christian cosmology. Wippel, "Condemnations of 1270 and 1277," 176. This same understanding was deduced also from Siger's writings on humankind. Wippel, "The Parisian Condemnations," 66. The error, however, was more closely associated with Boethius of Dacia. Wippel, "The Parisian Condemnations," 69.

100. On the exaggeration of the 219 propositions, see Thijssen, *Censure and Heresy*, 48–49, and Wippel, "The Parisian Condemnations" 71.

101. Mandonnet has identified forty articles (180–219) dealing with errors in theology. Mandonnet, *Siger de Brabant et l'averroïsme latin au XIIIme siècle*, II, 175–91.

102. Hissette, *Enquête sur les 219 articles*, 316–18.

103. The emphasis is Thijssen's, *Censure and Heresy*, 41.

104. Bacon, *Opus minus*, 323.

105. Burr, *Olivi's Peacable Kingdom*, 29.

106. "*ibi ei invenitur physica, que est c[oron]a salutis eterne. Ibi logica que est elisio dyabolice fraudis, ibi etica que est exemplar et speculum toius scientatis. Ibi est sciencia naturalis que uera passio, Ibi mathematica in forma crucis.*" MS Assisi 66, 63r; MS Assisi 71, 28r.

107. MS Assisi 66, 66r–64v; MS Assisi 71, 28r.

108. "[72.] That celestial bodies have eternity of substance but not eternity of motion. [76.] That a motive intelligence of the heavens influences the rational soul just as a celestial body influences a human body. [78.] That nothing could be new unless the sky were varied with respect to the matter of generable things. [104.] That from different [zodiacal] signs of the sky diverse conditions are assigned in men, both in respect to spiritual gifts and temporal things. [107.] That God could not have made prime matter without the mediation of a celestial body. [111.] That the elements have been made in a previous generation from chaos; but they are eternal. [112.] That the elements are eternal. However, they have been made [or created] anew in the relationship in which they have

now. [121.] That no form coming from outside can become one with matter. For what is separable does not make [or become] one with what is corruptible. [154.] That our will is subject to the power of celestial bodies. [156.] That the effects of stars on free will are hidden. [188.] That it is not true that something could be made from nothing, and also not true that it was made in the first creation." The English translations of these propositions have been taken from Edward Grant's translation, Grant, *A Source Book in Medieval Science*, 48–50. N.B., Grant uses the numbering system from Denifle and Chatelain, *Chartularium Universitatis Parisiensis* I, 543–55. This order antedates that of Mandonnet, and I have taken the liberty of altering the numeration (and reordering the propositions) to coincide with Mandonnet's. A concordance of the various numbering systems is available in Hissette, *Enquête*, 319. I raise this issue again in more detail in Chapter 4.

109. Rupescissa, *De consideratione*, 18. Vitalis, *Pro conservanda*, 9.

110. Discussion of Olivi's understanding of poverty can be found in Lambert, *Franciscan Poverty*, 160–61, and Burr, *The Spiritual Franciscans*, 50–55, 57–65.

111. See the sustained discussion in Burr, *Olivi's Peaceable Kingdom*, 132–56.

112. Burr, *The Spiritual Franciscans*, 262–63.

113. This argument has been made by Lerner, and echoed by DeVun. Lerner, "Historical Introduction, 85; Lerner, *The Feast of Saint Abraham*, 74; DeVun, *Prophecy, Alchemy* 32–33, 48. On Rupescissa's view of the Spirituals, see Rupescissa, *Liber ostensor*, 610–50, esp. 634–38.

114. For a discussion of how Joachim's *viri spirituali* were transformed into *viri evangelici* through the influence of Olivi and the Franciscan Spirituals, see Reeves, *The Influence of Prophecy*, 209–10.

115. On the meaning(s) of poverty among the friars, see Lambert, *Franciscan Poverty*, 33–72, 133–95; Burr, *The Spiritual Franciscans*, 50–65. John conflates the enemies of the Church with enemies of his understanding of "evangelical poverty (*evangelica paupertate*)." Rupescissa, *Liber ostensor*, 574.

116. Burr, *The Spiritual Franciscans*, 78.

117. Burr, *Olivi's Peaceable Kingdom*, 95–96.

118. John describes them as a Franciscan remnant in LSE, 151–2. In the *Liber ostensor*, however, he predicts that all the religious orders will suffer oppression, and that "all the bloated monasteries . . . will be devoured (*omnia monasteria pinguissima . . . devorabuntur*)." Rupescissa, *Liber ostensor*, 565–66. John's description of the oppression of the Friars Minor relies in this text on the vision of Jacob of Massa, an intimate of John of Parma, the Minister General and rigorist disgraced in the scandal of the Eternal Gospel. Rupescissa, *Liber ostensor*, 567–68. In the *Liber ostensor*, John calls Michael of Cesena a "*pseudopastor*" and "rapacious wolf (*lupus rapax*)," among other unsavory descriptions. Rupescissa, *Liber ostensor*, 636.

119. John notes that the Franciscans would be "divided into three different groups (*dividi in tres partes diversas*)". The first of these groups was obviously made up of conventual Franciscans, who "pant with infernal ambition to be promoted to the highest ecclesial offices and to be raised up with worldly riches (*luciferana ambitione anhelant ad summas*

ecclesiasticas dignitates promoveri et divitiis seculi)." Ultimately, "all will fall, with all of their progeny cleaving to them in laxness, into the heretical bosom of a false pope (*omnes cadent, cum omnibus suis filiis in laxationibus adherentibus eis, in hereticale gremium falsi pape*)." Rupescissa, LSE, 151. The second group are the Michaelists, in that they are of "the disposition of the Ghibellines (*de affectione populi gibelini*)," namely those who had once sought refuge in the court of Louis of Bavaria. Rupescissa, LSE, 152. Finally the third group, which is made up not only of Franciscans, but all of those who "try to observe the rule [of Francis] (*conantur regulam observare*)." This group holds to the pope and the integrity of the Roman Church. Rupescissa, LSE, 151–2. It is this remnant of true friars who hold to the pope who will "be sent by the pope to convert the world (*mittendi a papa ad seculum convertendum*)" and will be so many that they "cannot be counted (*numerari non possunt*)," all of them "living in such poverty and separation from the world that any one of them will seem like a new Francis in holiness (*in tanta paupertate et abstractione et separatione ab hoc mundo viventes ut quilibet in sanctitate quasi alter videatur Franciscus*)." Rupescissa, LSE, 202–3. For a similar passage on the future role of the Franciscans, see also Rupescissa, *Liber ostensor*, III.15.

120. "God made provision against the heresy of the leaderless, such as the aforesaid John XXII . . . (*contra heresiam acephalistarum providit Deus ut prefatus papa Johannes . . .*)" *Liber ostensor*, 586. John uses the term *acephalistae* in the *Liber ostensor* to refer to the Michaelists, and Michael himself is an *acephalista*. Rupescissa, *Liber ostensor*, 585, 636.

121. Although the argument John makes here might be taken universally, he singles out the Michaelists in a later passage. Rupescissa, *Liber ostensor*, 586–91.

122. "*nullus pauper evangelicus ab unitate et obediencia prelatorum, potissime pape et cardinalium, separare se debet, sed a contaminatione universorum scelerum.*" Rupescissa, *Liber ostensor*, 585.

123. "*evangelici pauperes, paupertatis altissime professores et defensores qui ab obediencia et unitate pape et cardinalium nequaquam recesserunt a papa Johanne XXII.*" Rupescissa, *Liber ostensor*, 585.

124. Rupescissa, *Liber ostensor*, 581, 622.

125. Lambert, *Franciscan Poverty*, 133. Lambert argues that Bonaventure's academic definition of poverty in his *Apologia pauperum* (Bonaventura, *Opera omnia*, VIII, 233–330) marked a turning point in the Franciscan understanding of poverty (and the understanding of Franciscan poverty). Notably, Bonaventure's *apologia* defined the key terms of the poverty dispute—dominion and *usus*. Lambert's discussion of the *apologia* is in *Franciscan Poverty*, 133–39.

126. Peter Olivi, on the other hand, is not a target of John's. He vigorously defends Olivi, stating that "during his entire life he forbade any of the brothers from leaving the community, and said of Brother Angelo and his brothers after their severing ties from the community, 'The smoke of their pride rises forever' (Apoc 14:11). (*toto tempore vite sue prohibuit ne aliquis fratrum a communitate recederet, et dixit de fratre Angelo et fratribus ejus post separationem eorum a communitate:* Fumus superbie eorum ascendet in secula seculorum.)" Rupescissa, *Liber ostensor*, 636.

127. "*usque hodie . . . sine capite generali ministro et sine obediencia et sine corectore.*" Given the fact that John must have known that Angelo was dead, we can assume his use of "*usque hodie*" is an oblique reference to Angelo's still living followers. Rupescissa, *Liber ostensor*, 634.

128. Rupescissa, *Liber ostensor*, 635.

129. "*non contentus de Regula fratrum minorum, quamdam compilationem composuit ex omnibus regulis, quasi esset eam perfectius aut tutius servare quam Regulam seraphici viri patris nostri angelici Franciscis.*" Rupescissa, *Liber ostensor*, 638.

130. In 1334, John of Valle won permission from Guirel Ot for himself and a group of four companions to strictly observe the Franciscan Rule, offering an (erstwhile) orthodox alternative to the Fraticelli in Umbria. Gentile of Spoleto, one of John of Valle's companions, and his followers were prosecuted, however, for " 'spiritual' tendencies (*tendances 'spirituelles'*)" (Rupescissa, *Liber ostensor*, 518n.), and the community lost their authorization in 1355. In 1368, the Observant movement again found traction under the leadership of Paoluccio dei Trinci. Burr, *The Spiritual Franciscans*, 303–4. John's principal antagonist within the Franciscan Order, Guillaume Farinier (d. 1361), was also Gentile's primary opponent. Rupescissa, *Liber ostensor*, 518n. Though John of Valle's experiment is quite significant when viewed through the lens of history, it may not have seemed so at the time – or at least not enough to draw John's prophetic eye. One can only speculate as to what John might have made of the Observants, and whether he would have supported their obedience to the papacy and strict life, or have castigated them for sowing further division.

131. Lerner, "Historical Introduction," 20–21.

132. Rupescissa did not identify the Avignon popes as antichrists. Rather, he gave that title to Louis of Bavaria, protector of the Michaelists and his antipope, Nicholas V, as well as other opponents of the Avignon papacy and French monarchy, such as Louis of Sicily. See Lerner, "Historical Introduction," 53–63.

133. This is the subject of the eighth chapter of the first book. Rupescissa, *De consideratione*, 36–38.

134. "*ad solvendam gravem inopiam et paupertatem futuram populi sancti et electi Dei.*" Rupescissa, *Liber lucis*, 121.

135. One of the suggested uses of the quintessence is to fortify soldiers before battle with the enemies of the Church. Rupescissa, *De consideratione*, 133–34. Compare Rupescissa, *Il libro della luce*, 121.

136. "*integrebitur ex omnibus fratris, prelatis vel subditis.*" Rupescissa, LSE, 151.

137. This does not mean, however, that John does not rehearse other arguments for poverty, especially those put forth in the bull *Exiit qui seminat* (1279). See, for example, his discussion in Rupescissa, *Liber ostensor*, 584–55.

138. Rupescissa, LSE, 202–3.

139. Rupescissa, LSE, 202–3.

140. Rupescissa, *Liber lucis*, 121.

141. Cf. DeVun, *Prophecy, Alchemy*, 49–51.

142. "*ad seruandum in toto tempore bellorum & tribulationum, & maxime tempore Antichristi,*" Rupescissa, *De consideratione*, 55.

143. Rupescissa, *De consideratione*, 122–23.

144. Rupescissa, *De consideratione*, 133–34.

145. DeVun, *Prophecy, Alchemy*, 102.

146. DeVun, *Prophecy, Alchemy*, 128.

CHAPTER 4. A SUBJUNCTIVE SCIENCE

1. For the importance of studying alchemical praxis, see Nummedal, "Words and Works in the History of Alchemy," 330–37.

2. P. Smith, "Secrets and Craft Knowledge in Early Modern Europe," 49–50.

3. J. Smith, "The Bare Facts of Ritual," 57–64.

4. Theissen, "The Attraction of Alchemy for Monks and Friars," 47. Bihl, "Statuta Glia Ordinis 1260, 1279 et 1292," 36.

5. Bihl, "Statuta Glia Ordinis," 36. Also see the statute in appendix 1 of de Dmitrewski, "Fr. Bernard Délicieux," 21–22. Bihl believes this statute, labeled 1313 by de Dmitrewski, actually belongs to a slightly earlier period. Bihl, "Statuta Glia Ordinis," 36.

6. "*operationes alchimiae, necromantiae, sortilegiorum vel superstitionum quarumlibet seu maleficiorum ac alias quascumque operationes doctrinarum vel artium suspectarum, quae in publico non docentur aut ab Ecclesia reprobatae sunt, et generaliter quascumque operationes praestigiosas et odibiles, sicuti sunt invocationes daemonum, incantantiones rerum et personarum.*" Bihl, "Statuta Glia Ordinis," 35. Cf. Dmitrewski, "Fr. Bernard Délicieux," 21.

7. Bihl, "Statuta Glia Ordinis," 35. Cf. Dmitrewski, "Fr. Bernard Délicieux," 21.

8. Bihl, "Statuta Glia Ordinis," 36. Dmitrewski, "Fr. Bernard Délicieux," 22.

9. Little, *Opus tertium*, 18.

10. For the treatments of *mathesis* and *matesis* before Bacon, see Marrone, *A History of Science, Magic, and Belief*, 134, 143–44; Peters, *The Magician, the Witch, and the Law*, 87.

11. "*nunquam veri philosophantes curaverunt de demonum invocatione, sed magici insani et maledicti.*" Little, *Opus tertium*, 48.

12. "*Sed ulterius procedit demencia matematicorum falsorum sine apparatione demonum.*" Bacon, *Tractatus brevis*, 6.

13. See Williams, "Roger Bacon and the *Secret of Secrets*," 374–75.

14. Little, *Opus tertium*, 48–49.

15. Bacon, *Tractatus brevis*, 6–7.

16. "*sine peccato incantare,*" Rupescissa, *De consideratione*, 54. "*erunt homines videntes naturaliter, vt incantanti,*" Rupescissa, *De consideratione*, 56.

17. Rupescissa, *De consideratione*, 35–36.

18. For instance, see John's discussion of extracting quintessence from silver and antimony. Rupescissa, *De consideratione*, 97, 103.

19. "*creavit Deus mirabilia in hoc mundo.*" Rupescissa, *De consideratione*, 24.

20. "*facies operationem Dei recte, et in eius virtute miracula super terram.*" Rupescissa, *De consideratione*, 28.

21. John argues that miracles come from the flow ("*influentias miras*") of the quintessence from heavenly bodies, especially the sun. Rupescissa, *De consideratione*, 22.

22. "*Et sicut carpentarius cum dolabra in manu fabricans arcas, non minus fabricat eas quam si absque dolabra fabricaret: sic Deus minus gubernat mundum: sic dedit talem influxum stellis, ut influant in rebus sicut et quomodo ipse vult, et non ultra.*" Rupescissa, *De consideratione*, 26.

23. Grant, *Planets, Stars, and Orbs*, 661–62. Grant points out that, unlike motion or light, influences were invisible.

24. Rupescissa, *De consideratione*, 25–26.

25. Fowden, *The Egyptian Hermes*, 76–79.

26. Fowden, *The Egyptian Hermes*, 122.

27. Patristic sources are conflicted over the "Christian" character of these texts. Augustine of Hippo derides them in *City of God*, while a number of Greek sources see the *Corpus Hermeticum* at least as akin to the Sybils—pagan prophecies of Christianity that had edifying properties.

28. Cosmic sympathy is an idea that appears to be derived from—and is certainly related to—antique Stoic ideas, especially those expounded by Posidonius. The simplest explanation of sympathy is that like produces like. The Stoic understanding of sympathy emerges from the doctrine of a living cosmos (itself derived from Platonic sources), where the creative energy of the divine flows through the universe at all times. This is distinct from earlier ideas of a prime mover, where the divine was only indirectly linked to the cosmos after the moment of creation. The Hermetic understanding of this idea follows. Fowden, *The Egyptian Hermes*, 77–78.

29. The *Asclepius* was, like virtually all the Hermetica, originally composed in Greek. The Hermetic lodges responsible for authoring and copying these texts wrote about them as though they were translations of either Coptic or Demotic in order to underscore their sacral characteristics. See Fowden, *The Egyptian Hermes*, 37–38.

30. The first text, called the *Poimandres*, is a revelatory dialogue between Poimandres (God) and Hermes. Subsequent dialogues are to be understood in the context of Hermes's *gnosis*. Copenhaver, *Hermetica*, 1–7.

31. Copenhaver, *Hermetica*, 2.

32. Copenhaver, *Hermetica*, 5–7.

33. Fowden, *The Egyptian Hermes*, 78.

34. Brown, *The World of Late Antiquity*, 102–3, quoted in J. Smith, "The Temple and the Magician," 186–87.

35. *Compendium aureum*, 213- 33. A good recent discussion of the textual history of the Latin Cyranides can be found in Voigts, "Plants and Planets," 37–40. Also useful is Thorndike, *A History of Magic and Experimental Science*, II, 233–35, as well as Delatte's introduction to his edition of the text. *Compendium aureum*, 209–11.

36. "One who reckons or counts" is quite close to the title of ancient Egyptian priests/time-keepers. A significant Egyptian priestly function was the keeping of ritual

time, which was, at least in part, an astronomical pursuit. See Campion, *The Dawn of Astrology*, 99–100. It is not clear if the title is meant to consciously remind the reader of connections to Egyptian priestly practice or is more of a generic. It is unlikely, however, that medieval readers would have made the connection to Egypt, especially given the fact that the translator simply transcribes the Greek rather than using the Latin *astrologus* or perhaps the closer synonym, *mathematicus*. This goes some way, I think, to proving the translator's lack of familiarity with the Hermetic context. *Compendium aureum*, 213.

37. *Compendium aureum*, 213–14.

38. *Compendium aureum*, 220, 26.

39. Smith and Hawthorne, *Mappae clavicula*, 3. Bischoff, "Die Überlieferung der technischen Literatur," 280–88. There is no critical edition of the text, but the twelfth-century Phillipps-Corning and the tenth-century Sélestat manuscripts have enough similarities that, for my purposes, I will write cautiously of the *Mappae clavicula* in general terms. I do not wish to downplay the differences, which are both important and instructive, in the various *Mappae clavicula* manuscripts. The variances present a picture, albeit an imperfect one, of the various advances in medieval techniques, but an in-depth discussion of this topic is beyond the scope of this project. For a thorough treatment of the manuscript history, see Smith and Hawthorne, *Mappae clavicula*, 3–13, Bischoff, "Die Überlieferung der technischen Literatur" 284–89, Halleux and Meyvaert, "Les origines de la *Mappae clavicula*," 7–10. On the translation of *Mappae clavicula*, see Halleux and Meyvaert, "Les origines de la *Mappae clavicula*," 11.

40. Smith and Hawthorne, *Mappae clavicula*, 4, 14. Bischoff, "Die Überlieferung der technischen Literatur," 284.

41. Smith and Hawthorne, *Mappae clavicula*, 14–15. Bischoff, "Die Überlieferung der technischen Literatur," 280. Halleux and Meyvaert, "Les origines de la *Mappae clavicula*," 7–8, 13, 18.

42. These eleven chapters likely constitute "a separately compiled group of contemporary recipes prefixed by the twelfth-century scribe to the longer manuscript," as Smith and Hawthorne suggest. Smith and Hawthorne, *Mappae clavicula*, 26n. It is quite an orderly, if brief, treatise, complete with its own verse prologue. While it supports the assessment of the *Mappae clavicula* as a compilatory text, it belies claims of disorganization, as discussed below.' "

43. Halleux and Meyvaert, "Les origines de la *Mappae clavicula*"; Bischoff, "Die Überlieferung der technischen Literatur."

44. Halleux and Meyvaert, "Les origines de la *Mappae clavicula*," 11–13.

45. "Die Wendungen '*sacri libri*,' (mit denen ja nicht die Bibel gemeint ist), '*heresis fatali munere concessa*,' '*per magnum deum*' scheinen mir keinen christlichen Tenor zu besitzen, und das führt mich dazu, eine Konjektur zu den ertsen Worten des Prologs vorzutragen: der Kommentar ist bestimmt:

Multis et mirabilibus in haec mei libri conscriptis
(so nach der Schlettstäder Handschrift) oder:

Multis et mirabilibus in hec meis conscriptis libris
(so nach dem Druck von Phillips);

haec mei oder *hec meis* ist zeifellos verdebt. Ich nehme an, daß *Hermaeis* oder *Hermis* zu lesen ist, und daß die ursprüngliche griechische Rezeptsammlung sich als eine Erklärung zu den Lehren des Hermes Trismegistos ausgab, der ja also einer der Erfinder der Alchemies gilt." Bischoff, "Die Überlieferung der technischen Literatur," 288–89. Halleux and Meyvaert, "Les origines de la *Mappae clavicula*," 14.

46. *Compendium aureum*, 215.

47. *Compendium aureum*, 218, 220, 221, 230.

48. Thorndike, *A History of Magic and Experimental Science*, II, 234.

49. The manuscript in question is Houghton MS Lat 220 ff. 36–39 as *Compendium medicine*, This manuscript largely conforms to the DeLatte edition, though Voigts has noted that the family of *Compendium* texts has a number of variant readings. Voigts, "Plants and Planets," 38. "*Pionia vocatur cruciara cuius radices etiam inveniuntur ad modum crucis et hec est verissima et habet enim radix illa proprietatem et virtutem ad demones expellendos.*" *Compendium medicine*, f. 37r.

50. For background on this text and its place in the antique world, see Moyer, "Thessalos of Tralles and Cultural Exchange," 39–56 and J. Smith, "The Temple and the Magician," 172–89. My summary of the narrative relies on a Latin form of the text. (pseudo-)Thessalos, *Thessali medici de virtutibus herbarum ad Claudium vel Neronem*, 254–57.

51. J. Smith, "The Temple and the Magician," 180–81.

52. For the description of the whole manuscript, see MS 9 in Moneti, Muzzioli, Rossi, Zamboni, *Catalogo dei Manoscritti della Biblioteca Casanatense*. The MS itself is cited henceforth as *Mirabilis cura uulnerum*.

53. *Mirabilis cura uulnerum*, col. 81b. Evidence for these three brothers as Franciscan comes from the fact that the manuscript itself is largely concerned with Franciscan letter forms, as well as the fact that the brothers are encountered while walking on the road seeking cures for wounds. One presumes these lesions have been incurred by the combination of their perambulations and poor footwear.

54. "*Tres boni viri fratres,*" *Mirabilis cura uulnerum*, col. 81b.

55. *Mirabilis cura uulnerum*, col. 82a.

56. "*Sanguis inde emanvit, non doluit, non rancavit, non fisutlavit, nec guttam nec ardorem nec putredinem fecit.*" *Mirabilis cura uulnerum*, col. 82a.

57. *Mirabilis cura uulnerum*, col. 82b–84a.

58. "God of the beloved saints, You who straighten the path and the way of the just, direct to me Raphael, the angel of your peace so that he might be joined to me and well-disposed. And through the tetragrammaton which Abraham carried on his brow and through Our Lord Jesus Christ your son who lives with and reigns with you forever and ever let no enemy snatch away this path from me. Amen. (*Dilectionum deus sanctorum qui dirigis item et viam iustorum, dirige mihi angelum tuum pacis raphaelem ut sit mihi commitatus et iocundus et ut nullus mihi uiam subripiat inimicus per tetragramaton quod in fronte*

portavit abraham et per dominum nostrum iesum christum filium tuum qui tecum vivit et regnat in secula seculorum amen.)" *Mirabilis cura uulnerum*, col. 84a.

59. "*nunquam etiam negare alicui uolenti scire facere predicta*," *Mirabilis cura uulnerum*, col. 84a.

60. Here I follow Macy, *Treasures of the Storeroom*, 36–50.

61. Bell, *Ritual Theory*, 90.

62. Bell, *Ritual Theory*, 7–8.

63. Bell, *Ritual Theory*, 93.

64. The office is found in Habig, *Omnibus*, 140.

65. Habig, *Omnibus*, 141–44.

66. Seligman, Weller, Puett, and Simon, *Ritual and its Consequences*, 22–32.

67. "*Unde brutum unitur solum accidentibus; infidelis nihil penitus credens superaddit quandam aptitudinem ad sacramentalem vel spiritualem manducationem; habens autem fidem informem superaddit quod unitur cognito, habens formatum credito, in quantum habet caritatem unitur ut dilecto.*" Guillelmi de Militona (William of Middleton), *Quaestiones*, 701.

68. Williams, *The Secret of Secrets*, 193.

69. "*Vidi enim medicum peritissimum qui majorem principem in regno Francie post regem, avarum, pusillaminem, tristem, malencolicum, debilem, et multis aliis viciis anime et corporis gravatum, curavit ab omnibus malis per hanc medicinam; et factus est largissimus, audacissimus, letissimus, et curatus ab omnibus passionibus melancolicis, et ab omnibus viciis corporis et anime naturalibus et moralibus.*" Bacon, *Secretum secretorum*, 105. While Bacon did not author the *Secretum secretorum*, the idiosyncrasy of his translation is such that I will cite it as his.

70. Williams, *The Secret of Secrets*, 193. Bacon, *Secretum secretorum*, 105–6.

71. Burke, *Opus majus*, 585–86.

72. Bonaventure, *Collationes in Hexaemeron*, XIII, 12, 390. See the discussion in Solignac, *La théologie symbolique*, 33–34.

73. Rupescissa, *De consderatione*, 112.

74. Williams, *The Secret of Secrets*, 313–14.

75. "*Unde una vetula paupercula suis precibus et meritis, bonitate Dei favente, potest mutare ordinem naturam.*" Bacon, *Tractatus brevis*, 4.

76. Bacon, *Opus minus*, 373–74.

77. See Bacon, *Epistolae Rogerii Baconis de secretis operibus artis et naturae*, 944–45.

78. Burke, *Opus majus*, 410–11. See also Delaurenti, *La puissance des mots*, 157–200.

79. "*circa cultum divinum statuerunt oraciones et sacrificia.*" Bacon, *Tractatus brevis*, 8.

80. Bacon, *Secretum secretorum*, 61–62.

81. "*astronomi Christiani debent pia devocione uti oracionibus ad Deum et sanctos, et eos invocare in omnibus operibus suis in auxilium.*" Bacon, *Tractatus brevis*, 8.

82. Bacon, *Tractatus brevis*, 8.

83. "*debent et possunt quedam opera facere conveniencia, ut facilius et melius accidant que intendat.*" Bacon, *Tractatus brevis*, 8.

84. "*coluerunt secundum graciam eis datam.*" Bacon, *Tractatus brevis*, 8.

85. Rupescissa, *De consideratione*, 20.

86. Rupescissa, *De consideratione*, 13.

87. Pereira, "L'Alchimista come medico perfetto," 77–79, 86–90.

88. Henceforth cited as Rupescissa, LSE.

89. John refuses the mantle of prophet. Rupescissa, LSE, 149, 151. Robert Lerner in his "Historical Introduction" to the LSE provides a thorough discussion of the magnitude of this distinction in the question of John's orthodoxy as well as its connection to Franciscan Joachism. Lerner, "Historical Introduction," 36–50.

90. "*revelavit arcana multa Philosophiae mundanae*," Rupescissa, *De consideratione*, 13.

91. Rupescissa, *Liber lucis*, 121, 143.

92. This takes up most of the text. For an example, however, see Rupescissa, *De consideratione*, 38–42.

93. "just as the saints shall be able to continue the works of the life of Christ longer and with more vigor through this book, so too would base men be able to endure in evil by its wicked use. I, however much I am able, shall continue this book on account of the saints alone, and shall commend it to Jesus Christ to guard (*sicut sancti per hunc librum poterunt continuare opera vitae Christi diutius et vehementius, ita et reprobi possent perverso usu diutius perseverare in malo. Ego autem, quantum in me est, propter solos sanctos librum continuo, et ipsum custodire Iesu Christo commendo*)," Rupescissa, *De consideratione*, 14.

94. Rupescissa, *De consideratione*, 118.

95. Here I follow Calvet, "À la recherche de la médicine universelle," 183.

96. "*une exégèse biblique du régime alchimique.*" Calvet, "À la recherche de la médicine universelle," 201.

97. On this point, see also Obrist, "Alchimie et allégorie scripturaire au Moyen Age," 257.

98. "*sicut et quomodo ipse vult, et non ultra.*" Rupescissa, *De consideratione*, 26.

99. "*Sed dices, cum omnia quae sunt corporalia in hoc mundo, et ad usum corporis, sint elementa vel ex elementis composita, ergo radix vitae non potest ab hominibus in hoc mundo vel seculo inveniri.*" Rupescissa, *De consideratione*, 18–19.

100. Rupescissa, *Liber ostensor*, 802.

101. "*iudicetur utrum Deus omnipotens ipse sit, qui michi aperuit intellectum . . . aut si deceptus fui per spiritum immundum.*" Rupescissa, LSE, 212. Reference to this passage from Lerner, "Historical Introduction," 37.

102. "*ergo phantasticum est laborare, ut in hac mortali vita quaeratur talis res, quae possit corpus nostrum ex mortali immortale reddere. . . . Phantasticum ergo esset dicere, ut Deus daret Adae extra paradisum aliquam rem per quam posset vivere in aeternum, ex quo ipsum eiecit, ne capiens de fructu ligni vitae viveret in aeternum. . . . Ergo subsidium illud vitae ad terminum prolongandum esset phantasticum et inane.*" Rupescissa, *De consideratione*, 16–17.

103. "*Ratio autem dictat, quod corpus corruptibile velle conservare per rem putridam et corruptibilem, et rem formare per rem deformati subiectam, et facere incorruptibile per rem*

deficientem, et infirmum sanare per infirmam, et turpe, pulchrum facere per rem foetidam, et debilem firmum per rem infirmam, est phantasticum et inane." Rupescissa, *De consideratione*, 17.

104. On Ockham, see Grant, *Planets, Stars, and Orbs*, 258–59.

105. "*phantastici reputantur.*" Rupescissa, *De consideratione*, 25.

106. "*Et omne remedium ab Ecclesia prohibitum, est euitandum.*" Rupescissa, *De consideratione*, 145.

107. Rupescissa, *De consideratione*, 163.

108. "*si curare praesumeres, phantasticum esset attentare pariter & insanum.*" Rupescissa, *De consideratione*, 164.

109. Rupescissa, *De consideratione*, 165.

110. "*perfecte curantur & vitantur cum quinta Essentia nostra: & talia vniversa pericula vitantur.*" Rupescissa, *De consideratione*, 166.

111. "*Remedium contra phantasticas passiones, imaginationes, et fatuas demonum infestationes et tentationes,*" Rupescissa, *De consideratione*, 131.

112. Rupescissa, *De consideratione*, 11–12, and "*divina benignate sanare: impedimenta sanctae orationi et meditationi mirabiliter effugare, tentationibus etiam daemonum,*" at Rupescissa, *De consideratione*, 12.

113. "*omnis spiritus angelicus, qui licet longum, latum & profundum corporales dimensiones non habet.*" Rupescissa, *De consideratione*, 137.

114. Rupescissa, *De consideratione*, 134–36.

115. John says that "our spirit after separation from the body [has] such a power of perception, not carnal, but spiritual, through which it is subjected to the passions (*spiritus noster post separationem a corpore habeat talem potentiam sensitiuam, non carnalem sed spiritualem, per quam subiicitur passionibus*)." He ascribes this idea to Augustine in "his book on the spirit and soul (*libro de spiritu et anima*)," which likely refers to discussions in *De Animae Quantitate* on the soul's awareness. Rupescissa, *De consideratione*, 136.

116. Rupescissa, *De consideratione*, 137.

117. "*quod manifesta experientia docet.*" Rupescissa, *De consideratione*, 138.

118. Many natural philosophers disdained pure empiricism, of course, in the same way scholastics balked at novelty, that is to say, reflexively. Nevertheless, alchemy required not only observation, but also operation, and in this way was distinct from many other branches of learning in the fourteenth century. See Crisciani, "Artefici sensati," 135–59.

119. The cure, however, was not achieved through the use of the quintessence, but rather the herb *hypericon*, or St. John's wort. John advises mixing St. John's wort with the quintessence for a more effective remedy. Rupescissa, *De consideratione*, 141.

120. Elliot, "The Physiology of Rapture," 141–75, esp. 159–60.

121. "*non solum quinta Essentia perfecte sanat, sed etiam aqua ardens cum aliquantulo eorum que choleram nigram epellunt, et choleram nigram superfluam purgant, splenem sanat, turbationem cogitationum tollit, cor triste laetificat, cerebrum et omnes eius potentias mundificat, laetas cogitationes inducit, artificium tentationis daemoniacae aufert, et imaginationes desperatiuas annullat, ac facit malorum quorunque oblivisci, et homines ad sensum naturalem reverti.*" Rupescissa, *De consideratione*, 60.

122. For some philological background on the term, see Du Cange, *Glossarium*, 204; Blaise, *Lexicon Latinitatis Medii Aevi*, 684; Diefenbach, *Glossarium Latino-Germanicorum* , 225; Grubmüller and Schnell, eds., *Vocabularius Ex quo*, 989.

123. Rupescissa, *De consideratione*, 162–63.

124. "*in horrere visionis nocturne . . . emisit bestiam nigram ab oriente.*" Rupescissa, *Liber ostensor*, 326. The parallel description is Rupescissa, *Liber ostensor*, 372.

125. Rupescissa, *De consideratione*, 165–66.

126. "*Reverendissime in Christo pater et domine domine Guillerme, sacrosanctae Romane Ecclesie cardinalis . . . ego frater Ioannes de Rupescissa ordinis Fratrum Minorum . . . ad mandatum vestrum descripsi seriem notabilium eventuum futurorum michi in carceribus apertorum, prout melius et verius potero recordari. Modus revelandi fuit iste: cum anno Domini nostri Iesu Christi MCCCXLV multis diebus flerem vinctus ferro, in carcere luti in conventu Figiaci, stupens et mirans quare cum tanta crudelitate missus essem per fratrem Guillermum Farinerii, tunc ministrum Aquitanie, in carcere in quo erat per totum vere lutum et molle, et quare permisisset Deus me tradi in manus crudelium dominorum ad vocem testium periurorum et falsorum, sicut dies Iudicii declarabit, fuit michi clare datum intelligi quod idea in tantam temptationem incideram quia futurum erat ut ego Antichristum et eius genus et terram, antequam appareret ut Antichristus, seculo revelarem et ut habilitarer ad intelligendum futura conclusus fueram in validissimas temptationes et afflictiones quas iam patior quinque annis. Quo intellecto augmentavi orationem et penitentiam aggravavi et vigilias ampliavi. Factum est autem eodem anno in mense iulii, circa festum sancti Iacobi Apostoli, cum essem vigilans et orans stans pedes in medio carceris, tenens virgam in manu, subito in instanti, in ictu oculi (1 Cor 15, 52) apertus est intellectus meus et de futuris intellexi visione intellectuali que consequenter summatim et breviter describuntur.*" Rupescissa, LSE, 137–38.

127. Lerner, "Historical Introduction," 38. I think this explanation suffices for the circumstances of the composition of the *Book of Hidden Events*, which John composed for his defense in a heresy trial. Some years later, however, John contradicts the notion that prophecy is closed. In the *Liber ostensor*, he says that prophecy "endures in the Church of Christ from Christ up to now and shall continue in the general Church without interruption up to the end of the world. (*duravit in Ecclesia Christi a Christo usque modo et continuabitur in Ecclesia generali sine interruptione usque ad finem mundi.*)" Rupescissa, *Liber ostensor*, 762–63. What ended, argues John, is the "mode of prophecy" which foretold the Son (a rather obvious assertion) and that which revealed new divine law. Rupescissa, *Liber ostensor*, 763.

128. "*quo martirium antiquitus.*" Rupescissa, *Liber ostensor*, 517.

129. "*in momento.*" Rupescissa, LSE, 146. This passage also echoes Joachim.

130. Rupescissa, LSE, 146, 164, 212.

131. "*pondus celestis dulcedinis et presentie glorie magne Dei . . . supervenit in me.*" Rupescissa, LSE, 212.

132. Robert Lerner describes this vision precisely as ecstasy. Lerner, "Historical Introduction," 38.

133. "*in visu apparuit michi angelus Domini.*" Rupescissa, *Liber ostensor*, 521.

134. For the Toulouse vision, see Rupescissa, *Liber ostensor*, 248–49.

135. Rupescissa, *Liber ostensor*, 108. Rupescissa, LSE, 213, 214.

136. Rupescissa, *Liber ostensor*, 736.

137. Rupescissa, *Liber ostensor*, 736–37.

138. Rupescissa, *Liber ostensor*, 737–39.

139. Rupescissa, *Liber ostensor*, 739–40.

140. Rupescissa, *Liber ostensor*, 741–42.

141. Rupescissa, *Liber ostensor*, 742.

142. Rupescissa, *Liber ostensor*, 750.

143. Rupescissa, *Liber ostensor*, 762–63.

144. "*usque modo et . . . sine interruptione usque ad finem mundi.*" Rupescissa, *Liber ostensor*, 762–63.

145. Rupescissa, *Liber ostensor*, 763.

146. "*Prima est futura previdere per Dei vel angeli sancti revelationem divinam in sompnis.*" Rupescissa, *Liber ostensor*, 764.

147. For instance, John describes a night vision in *Liber ostensor*, 326. At *Liber ostensor*, 372, he describes a vision coming while "in sleep (*in sompnis*)." Other sleeping visions occur in Rupescissa, *Liber ostensor*, 404, 413.

148. Rupescissa, *Liber ostensor*, 521.

149. "*audire futura vigilando, ex ore angelorum sanctorum.*" Rupescissa, *Liber ostensor*, 767.

150. "*Diferencia secunda est previdere futura per intelligenciam Scripturarum sanctarum et prophetarum . . . Hunc modum habuit Joachim et multi alii ipsum habuerunt post eum.*" Rupescissa, *Liber ostensor*, 764–65.

151. Rupescissa, *Liber ostensor*, 765–66.

152. "*prophetare per concordiam utriusque divini Testamenti.*" Rupescissa, *Liber ostensor*, 767.

153. "*nova Troia—pecatrice Avinione*," Rupescissa, *Liber ostensor*, 513.

154. Since John is working through Joachim (who works from the principle of intertestamental concordance), he also identifies various scriptural passages hinting at his arrival. For example, he cites passages from Daniel 12 (Rupescissa, *Liber ostensor*, 514–17, 529–30) and Romans 9 and 11 (Rupescissa, *Liber ostensor*, 515) among others.

155. "*fantasticum et delirum, amentem penitus et insanum.*" Rupescissa, *Liber ostensor*, 513.

156. Rupescissa, *Liber ostensor*, 517–32.

157. Rupescissa, *Liber ostensor*, 768–69.

158. For instance, there is penitence (Rupescissa, *Liber ostensor*, 769), followed by submitting one's will to that of Christ (Rupescissa, *Liber ostensor*, 770), to mystically "through the deepest contemplations (*per altissimas contemplationes*)" suffering the Passion (Rupescissa, *Liber ostensor*, 770), to exercising patience in the face of injury or attack (Rupescissa, *Liber ostensor*, 771), and using prayer to avoid "the temptation of Antichrist (*temptationem Antichristi*)" by immersing oneself "into the fire of the heart of Christ (*in*

ignem cordis Christi)," (Rupescissa, *Liber ostensor*, 771). Only the final two species of martyrdom have to do with physical suffering. (Rupescissa, *Liber ostensor*, 772–78).

159. "*ponit frater Petrus Johannis.*" Rupescissa, *Liber ostensor*, 778. For discussion of Olivi's view of visionaries, see Burr, *Olivi's Peaceable Kingdom*, 124–25.

160. It is a point of interest that John complains about enduring for years the insane and demoniacal ramblings of a fellow prisoner, Simon Legat, an English priest. Rupescissa, *Liber ostensor*, 523–25.

161. "*Primo non desidare visiones, raptus, et sompnia seu revelationes contra naturam, quia sancti communiter talia vitaverunt—quantum in eis erat—, quia talibus fantastice illusiones a demonibus immituntur. Secundo in oratione vel contemplatione non sustinere delectionem, quantumcumque magnam vel parvam, que te duceret in presumptionem et in tui ipsius vanam estimationem et superbiam. . . . Quinto, ut vites familiaritates personarum sequentium, amantium et laudantium dictas illusiones et fautores earum. . . . Sexto, ut habeas cautelam magnam, quando audies eorum visiones, sentimenta et raptus, an sint et ducant contra doctrinam Christi et sanctorum et Ecclesie ac Scripture sacre sane intellecte et contra bonos mores. . . . Et [si] apprehendas certitudinaliter dissonare, subito respue et impugna audacter! . . . . Octavo, si per visiones, revelationes, raptus et sentimenta moveris ad faciendum aliquid grave, arduum et insolitum opus, si dubitas rationabiliter an a Deo sit et sit acceptum et tibi utile, diferas illud opus agredi donec inspexeris universas circumstancias et maxime finales.*" Rupescissa, *Liber ostensor*, 779–82.

162. "*iudicetur utrum Deus omnipotens ipse sit, qui michi aperuit intellectum . . . aut si deceptus fui per spiritum immundum.*" Rupescissa, LSE, 212. Reference to this passage from Lerner, "Historical Introduction," 37.

163. Rupescissa, *Liber ostensor*, 526.

164. "*scito quod, de uberibus fere matris mee, sensii in me quamdam complexionem pestiferam a meis malis moribus . . . quod fui et sum adhuc pronissimus ad iram et motum irascendi pro nichilo, ex superflua superbia et inflatione cordis, et ad male dicendum et ad reprehendendum in aliis ea que in me ipso temperare non novi.*" Rupescissa, *Liber ostensor*, 526.

165. "*aborreo me ipsum, be[s]tiam sine ratione me videns, cum me videam sine causa et ratione irasci et passim sanctos sine necessitate jurare.*" Rupescissa, *Liber ostensor*, 526. We ought to take "*jurare*" here to mean swearing in the modern sense, that is, using foul language, or more akin to John's usage, taking the Lord's name in vain.

166. "*abundantia cholerae . . . communiter habet annexam alientionem sensus & apparitionem rerum phantasticarum.*" Rupescissa, *De consideratione*, 161.

167. Rupescissa, *De consideratione*, 131.

168. Rupescissa, *De consideratione*, 151 (for a chest cold); 145 (for sores caused by prison chains.)

169. Seligman et al., *Ritual and Its Consequences*, 30.

170. "*Anaxagoras antiquus philosophus simplicter loquendo prophetavit. Hector post mortem suam denunciavit destructionem Troie fore; dum dormiebat in lecto suo previdit eam. Itaque et novus Hector militans in Ecclesia nove Troie novam destructionem annunciabit, sed*

fantasticus reputabitur." Joachim of Fiore, *Liber de Flore*, 55 n50, quoted by Rupescissa, *Liber ostensor*, 509–10.

171. Emphasis mine. "*Ego autem non exposuissem hanc prophetiam Joachim que de me facta est si in ea aliquid laudis contineretur de me.*" Rupescissa, *Liber ostensor*, 533.

CONCLUSION

1. Rupecissa, *De consideratione*, 112.

2. Leong and Rankin, "Introduction," 8. Eamon, *Science and the Secrets of Nature*, 15–16.

3. Eamon, *Science and the Secrets of Nature*. For some discussion of the "naturalism" of the mirabilia, see Datson and Park, *Wonders and the Order of Nature, 1150–1750.*

4. Eamon, "How to Read a Book of Secrets," 26.

5. Steven J. Williams, *The Secret of Secrets*, 347.

6. Eamon, "How to Read a Book of Secrets," 28–29.

7. Eamon, "How to Read a Book of Secrets," 44.

8. For discussion of this text, see Thorndike, *A History of Magic and Experimental Science*, III, 628.

9. Bernard of Treves, *De chymico miraculo*, 6.

WORKS CITED

PRIMARY LITERATURE

Alexander Halensis, *Opera omnia*. Paris: Antonii Bertier, 1647.

Aristotle. *De Caelo*. Trans. J. L. Stocks. Oxford: Clarendon, 1922.

(pseudo-)Aristotle. *Secretum Secretorum*, in *Opera hactenus inedita Rogeri Baconi*, ed. Robert Steele, fasc. V. Oxford: Clarendon, 1920.

Armstrong, Regis J., J. A. Wayne Hellman, and William J. Short, eds. *Francis of Assisi: Early Documents*. 3 vols. New York: New City Press, 1999.

Augustine. "The Literal Meaning of Genesis." In Augustine of Hippo, *On Genesis*. Hyde Park, N.Y.: New City Press, 2002.

———. *Sancti Avreli Avgvstini De Genesi ad litteram libri dvodecim; eivsdem libri capitvla*, ed. Joseph Zycha. Vindobonae, F. Tempsky; Lipsiae, G. Freytag, 1894.

Avicenna (ibn Sina). *Avicennae De congelatione et conglutinatione lapidum*, ed. Eric John Holmyard and Desmond Christopher Mandeville. Paris: Guethner, 1927.

Bartholomaeus Anglicus. *Venerandi patris Bartholomei Anglici . . . de rerum proprietatibus* Nuremberg: Joannis Koberger, 1519.

"Benedict of Peterborough" (Roger of Hoveden). *Gesta regis Henrici Secundi Benedicti abbatis: The Chronicle of the Reigns of Henry II. and Richard I.* Ed. W. Stubbs. London: Longmans, Green, Reader, and Dyer, 1867.

Bernard of Treves. *De chymico miraculo*. Basel: Petrus Perna, 1583.

Bonaventure. *Apologia pauperum,* In Bonaventura, *Opera Omnia* VIII. Florence: Quaracchi, 1898.

———. *Breviloquium*. In Bonaventura, *Opera Omnia* V. Florence: Quaracchi, 1891.

———. *Collationes in Hexaemeron*. In Bonaventura, *Opera Omnia* V. Florence: Quaracchi, 1891.

———. *Commentary on the Sentences*. In Bonaventure, *Opera Omnia*, ed. A. C. Peltier. Paris: L. Vivès, 1864.

———. *Itinerarium mentis in Deum*. In Bonaventura, *Opera Omnia* V. Florence: Quaracchi, 1891.

———. *Saint Bonaventure's De Reductione Artium Ad Theologiam*. Ed. Emma Thérèse Healy. Saint Bonaventure, N.Y.: Saint Bonaventure College, 1940.

Bullarii Franciscani Epitome. Ed. C. Eubel. Apud Claras Aquas, 1908.

Burton, P. H., J. Balserak, H. A. G. Houghton, and D. C. Parker. "Vetus Latina Iohannes Synopsis 2.0." Vetus Latina Iohannes: The Verbum Project, The Old Latin Manuscripts of John's Gospel." April 30, 2015. Accessed May 19, 2015. http://www.iohannes.com/vetuslatina/index.html.

Chartularium Universitatis Parisiensis. Ed. H. Denifle and E. Chatelain. Paris: Delalain, 1889–1897.

Chronica XXIV Generalium Ordinis Minorum. In *Analecta Francescana* III. Florence: Quaracchi, 1897.

Clément V. "Deus scientiarum," in *Lettres communes*, Avinione 8 Sept. 1309, Reg. Vat. 57.

Compendium aureum. In Louis Delatte, ed., *Textes latines et vieux français relatifs aux Cyranides*. Liège: Faculté de Philosophie et Lettres, 1942.

Compendium medicine (*Compendium aureum*). Houghton Library, Ms. Lat. 220.

Constantine of Pisa. *The Book of the Secrets of Alchemy*. Critical ed., intro., trans., and commentary Barbara Obrist with Centre National de la Recherche Scientifique. Leiden: Brill, 1990.

Francis of Assisi. *The Writings of Saint Francis*. Trans. and ed. Paschal Robinson, O.F.M. Philadelphia: Dolphin Press, 1905.

Guillelmi de Militona (William of Middleton). *Quaestiones de sacramentis*. Florence: Quaracchi, 1961.

Habig, Marion A., ed. *St. Francis of Assisi: Writings and Early Biographies; English Omnibus of the Sources for the Life of St. Francis*. 3rd ed. Chicago: Franciscan Herald, 1973.

Hugh of St. Victor. *Hugonis de Sancto Victore Didascalicon de studio legendi*. Ed. Charles Henry Buttimer. Washington, D.C.: Catholic University Press, 1939.

Jean de Venette. *The Chronicle of Jean de Venette*. Ed. Richard A. Newhall, trans. Jean Birdsall. New York: Columbia University Press, 1953.

———. *Cronica*. London: British Library, MS Arundel 28.

Joachim of Fiore. *Expositio in Apocalypsim*. Venice, 1527; repr. Frankfurt: Minerva, 1964.

———. *Liber de Concordia Noui ac Veteris Testamenti*. Ed. E. Randolph Daniel. *Transactions of the American Philosophical Society* 73, 8 (1983).

———, *Liber de Flore*, ed. H. Grundmann, *Historisches Jahrbuch* 49 (1929).

John Bale, *The Actes of English Votaries or Unchast examples of the Englsih Votaryes*. London: J. Tysdale, 1560.

———. "The Examination and Death of Lord Cobham." In *Select Works of John Bale*. New York: Johnson Reprint, 1968.

———. "The Image of Both Churches." In *Select Works of John Bale*. New York: Johnson Reprint, 1968.

John XXII. *De Crimine Falsi*. In *Extravagantes Communes*, Lib. V, Tit. 6 in *Corpus Iuris Canonici*. Lipsiae, ex officina Bernhardi Tauchnitz, 1879–81. Tom. II.

———. *Prodiens*. In *Extravagantes Iohannis XXII*, ed. Jacqueline Tarrant. Vatican City: Biblioteca Apostolica Vaticana, 1983.

———. *Super illius specula*. In *Bullarium Romanum novissimum a B. Leone Magno, usque ad S. D. N. Urbanum VIII*. Rome: R. Camerae Apost., 1638.

John of Rupescissa. *De famulatu philosophia* (*De consideratione quintæ essentie*). Cambridge, Mass., Houghton Library, MS Lat 220.

———. *Ioannis de Rvpescissa qvi ante CCCXX. annos vixit: De consideratione quintæ essentie rerum omnium.* Basilae, 1561.

———. *Liber ostensor quod adesse festinant tempora.* Ed. André Vauchez, Clémence Thévenaz Modestin and Christine Morerod-Fattebert et al. Rome: École Française de Rome, 2005.

———. *Liber secretorum eventuum.* Ed. Robert E. Lerner and Christine Morerod-Fattebert. Fribourg: Éditions Universitaires, 1994.

———. *Il libro della luce.* Ed. Andrea Aromatico.Venezia: Marsilio, 1997.

Julian of Speyer. *Vita Sancti Francisci* in *Officio ritmico e Vita Secunda.* Ed. Vergilio Gamboso. Padova: Edizioni Messagero, 1985.

Mappae clavicula: A Little Key to the World of Medieval Techniques. Ed. Cyril Stanley Smith and John G. Hawthorne. Philadelphia: American Philosophical Society, 1974.

Michael Scot (?). *Ars Alchemie.* In Charles Homer Haskins, "The 'Alchemy' Ascribed to Michael Scot." *Isis* 10, 2 (1928).

——— (?), *Ars Alchemie.* In S. Harrison Thomson, "The Texts of Michael Scot's *Ars Alchemie.*" *Osiris* 5 (1938).

Mirabilis cura uulnerum etiam letalium et apostematum et phistolarum et etiam que euellit quadrellos. Rome, Biblioteca Casanatense, MS 9.

Nicholas of Lyra. *Postilla super totam bibliam.* Straßburg: Minerva, 1492.

Paul of Taranto. *Theorica et Practica.* Ed. and trans. William R. Newman in Newman, "The Summa Perfectionis and Late Medieval Alchemy: A Study of Chemical Traditions, Techniques and Theories in Thirteenth Century Italy," III. Ph.D. dissertation, Harvard University, 1986.

Peter Abelard, *Expositio in hexaemeron.* Ed. Mary Romig and David Luscombe in *Petri Abaelardi opera theologica* V. Turnhout: Brepols, 2004.

Peter of Auvergne, *Questions on Aristotle's* de Caelo. Ed. Griet Galle. Leuven: Leuven University Press, 2003.

Peter Lombard. *Sententiae in IV libris distinctae* Grottaferrata: Editiones Collegii S. Bonaventurae ad Claras Aquas, 1971–81.

Peter Olivi. *Peter of John Olivi on Genesis.* Ed. David Flood, O.F.M. St. Bonaventure, N.Y.: Franciscan Institute, 2007.

Plato. *Timaeus.* Trans. D. Zeyl. Indianapolis: Hackett, 2000.

Ptolemy. *Centiloquium.* MS Bayerische StaatsBibliothek (BSB) Hss Clm 275.

———. *Liber quadripartiti Ptholemei; Centiloquium [ejusdem]. Centiloquium Hermetis.* Venice, 1493.

(pseudo-) Robert Grosseteste. *Die philosophischen Werke des Robert Grosseteste, Bischofs von Lincoln.* Ed. L. Baur. Münster: Aschendorff, 1912.

Roger Bacon. *Compendium studii philosophiae* in *Opera quaedam hactenus inedita.* Ed. J. S. Brewer. London: Longman, Green, 1859.

———. *De erroribus medicorum* in *Opera hactenus inedita Rogeri Baconi,* fasc. IX. Ed. Robert Steele. Oxford: Clarendon, 1928.

———. *De multiplicatione specierum*. In *Roger Bacon's Philosophy of Nature, A Critical Edition, with English Translation, Introduction, and Notes of* De multiplicatione specierum *and* De speculis comburentibus, ed. David C. Lindberg. Oxford: Clarendon, 1983.

———. *Epistolae Rogerii Baconis de secretis operibus artis et naturae*. In *Theatrum Chemicum* V, ed. Lazarus Zetzner. Strasbourg, 1622.

———. *Frater Rogerus Bacon in libro sex scientiarum*. In Fratris Rogeri Bacon, *De retardatione accidentium senectutits cum aliis opusculis de rebus medicinalibus*, ed. A. G. Little and E. Withington. Oxford: Clarendon, 1928.

———. *Opus minus*. In *Opera quaedam hactenus inedita*, ed. J. S. Brewer. London: Longman, Green, 1859.

———. *Opus tertium*. In *Opera quaedam hactenus inedita*, ed. J. S. Brewer. London: Longman, Green, 1859.

———. *Part of the Opus tertium of Roger Bacon*. Ed. A. G. Little. Aberdeen: University Press, 1912.

———. *The Opus majus of Roger Bacon*. Ed. J. H. Bridges. Oxford: Clarendon, 1897.

———. *The Opus majus of Roger Bacon*. Trans. Robert Bell Burke. Philadelphia: University of Pennsylvania Press, 1928.

———. *Secretum secretorum*. In *Opera hactenus inedita Rogeri Baconi*, ed. Robert Steele. Oxford: Clarendon, 1905.

———. *Tractatus Brevis*. In *Opera hactenus inedita Rogeri Baconi*, fasc. V, ed. Robert Steele. Oxford: Clarendon, 1905.

Roger Hoveden, *Chronica*, ed. W. Stubbs. London: Longman, 1868–71.

Salimbene de Adam. *Cronica*. Ed. Giuseppe Scalia. Bari: Laterza, 1966.

Scripta Leonis, Rufini et Angelo Sociorum S. Francisci. Ed. Rosalind Brooke. Oxford: Clarendon, 1990.

Stephen Tempier. "Condemnations of 219 Propositions." Trans. Ernest L. Fortin and Peter D. O'Neill. In Ralph Lerner and Muhsin Mahdi, eds., *Medieval Political Philosophy*. Ithaca, N.Y.: Cornell University Press, 1972.

(pseudo-)Thessalos of Tralles. *Thessali medici de virtutibus herbarum ad Claudium vel Neronem*. In Franz Cumont, ed., *Catalogus codicum astrologum graecorum*, VIII, 4. Brussels: Henri Lamertin, 1921.

Thomas Aquinas. *In Aristotelis libros De caelo et mundo de generatione et corruptione meteorologicorum expositio*. Ed. Raymundo Spiazzi, O.P. Torino: Marietti, 1952.

———. *Opera omnia* 31. Paris: Apud Ludovicum Vivès, 1871–1880.

———. *Opera omnia* 23. Parma: P. Fiaccadori, 1852–1873.

Thomas of Celano. *St. Francis of Assisi: First and Second Life of St. Francis*. Ed. and trans. Placid Hermann, O.F.M. Quincy, Ill.: Franciscan Press, 1988.

Vitalis of Furno (?). *Pro conservanda sanitate, tuendaque prospera valetudine, ad totius humani corporis morbos et aegritudines*. Mainz: Schoeffer, 1531.

——— (?). *Super Apocalypsim*. Assisi: Biblioteca del Sacro Convento, MS Assisi 66.

SECONDARY LITERATURE

Agrimi, Jole and Chiara Crisciani, "Savoir médical et anthropologie religieuse: Les représentations et les fonctions de la vetula (XIIIe–XVe siècle)." *Annales* 48, 5 (1993).

Anderson, C. Colt. *A Call to Piety: St. Bonaventure's* Collations on the Six Days. Quincy, Ill: Franciscan Press, 2002.

Bailey, Michael. *Magic and Superstition in Europe*. Plymouth: Rowan and Littlefield, 2007.

Bartlett, Robert. *The Natural and Supernatural in the Middle Ages*. New York: Cambridge University Press, 2008.

Bell, Catherine. *Ritual Theory, Ritual Practice*. New York: Oxford University Press, 1992.

Berthelot, Marcelin. *La chimie au moyen age*. Paris: Imprimerie Nationale, 1893.

Bianchi, Luca. "1277: A Turning Point in Medieval Philosophy?" In Jan A. Aersten and Andreas Speer, eds., *Was ist Philosophie im Mittelalter? Qu'est-ce que la philosophie au Moyen Âge? What is Philosophy in the Middle Ages? Akten des X. Internationalen Kongresses für mittelalterliche Philosophie der Société Internationale pour l'Étude de la Philosophie Médiévale 25. bis 30. August 1997 in Erfurt*. New York: Walter de Gruyter, 1998.

———. *Censure et liberté intellectuelle à l'université de Paris: XIIIe–XIVe siècles*. Paris: Belles Lettres, 1999.

Bigalli, Davide. *I Tartari e L'Apocalisse: Ricerche sull'escatologia in Adama Marsh e Ruggero Bacone*. Firenze: Nuova Italia Editrice, 1971.

Bihl, P. Michael. "Statuta Glia Ordinis 1260, 1279 et 1292." *Archivum Franciscanum* 34 (1941).

Bischoff, Bernhard. "Die Überlieferung der technischen Literatur." *Mittelalterliche Studien* 3 (1981).

Blaise, Albert. *Lexicon Latinitatis Medii Aevi: Praesertim ad Res Ecclesiasticus Investigandas Pertinens*. Turnhout: Brepols, 1975.

Blastic, Michael W. "Preaching in the Early Franciscan Movement." In Timothy J. Johnson, ed., *Franciscans and Preaching*. Boston: Brill, 2012.

Brenon, Anne. *Les femmes cathares*. Paris: Perrin, 1992.

Brown, Peter. *The World of Late Antiquity*. London: Thames and Hudson, 1971.

Boureau, Alain. *Satan the Heretic: The Birth of Demonology in the Medieval West*. Trans. T. Fagan. Chicago: University of Chicago Press, 2006.

Burr, David. "Antichrist and Islam in Medieval Franciscan Exegesis." In J. Tolan, ed., *Medieval Christian Perceptions of Islam*. New York: Garland, 1996.

———. *Olivi's Peaceable Kingdom*. Philadelphia: University of Pennsylvania Press, 1993.

———. *The Persecution of Peter Olivi*. Philadelphia: American Philosophical Society, 1976.

———. *The Spiritual Franciscans: From Protest to Persecution in the Century After Saint Francis*. University Park: Pennsylvania State University Press, 2001.

Bynum, Caroline Walker. *Christian Materiality: An Essay on Religion in Late Medieval Europe*. New York: Zone, 2011.

———. *The Resurrection of the Body in Western Christianity, 200–1336*. New York: Columbia University Press, 1995.

Calvet, Antoine. "À la recherche de la médicine universelle: Questions sur l'élixir et la thériaque au 14e siècle." In Chiara Crisciani e Agostino Paravicini Bagliani, eds., *Alchimia e medicina nel Medioevo*. Firenze: SISMEL Editzioni del Galluzzo, 2003.

Campion, Nicholas. *The Dawn of Astrology: A Cultural History of Western Astrology*. Vol. 1. New York: Continuum, 2008.

Capitanucci, Paolo. "Francescani e alchimia fra mito e realtà: la leggenda di Elia alchimista," In Paolo Capitanucci, ed., *Elia di Cortona tra realtà e mito: Atti dell'Incontro di Studio Cortona, 12–13 luglio 2013*. Spoleto: Centro Italiano di Studi sull'Alto Medioevo, 2013.

Chakrabarty, Dipesh. *Provincializing Europe: Postcolonial Thought and Historical Difference*. Princeton, N.J.: Princeton University Press, 2000.

Copenhaver, Brian P. *Hermetica: The Greek* Corpus Hermeticum *and the Latin* Asclepius *in a New English Translation, with Notes and Introduction*. New York: Cambridge University Press, 1992.

Crisciani, Chiara. "Artefici sensati: experientia e sensi in alchimia e chirugia (secc. XII–XIV)." In Chiara Crisciani and Agostino Paravicini Bagliani, ed., *Alchimia e medicina nel Medioevo*. Firenze: SISMEL, 2003.

———. *Il papa e l'alchimia, Felice V, Guglielmo Fabri e l'elixir*. Roma: Viella, 2002.

Crisciani, Chiara and Michela Pereira. *L'Arte del Sole e della Luna: Alchimia e filosofia nel medioevo*. Spoleto: Centro Italiano di Studi sull' Alto Medioevo, 1996.

Dahan, Gilbert. "Herméneutique et méthodes d'exégèse" In Gilbert Dahan, ed., *Nicolas de Lyre: Franciscain du XIVe* siècle: *Exégète et théologien*. Paris: Institut d'Études Augustiniennes, 2011.

Daniel, E. Randolph. *The Franciscan Concept of Mission in the High Middle Ages*. Lexington: University Press of Kentucky, 1975.

Datson, Lorraine and Katherine Park. *Wonders and the Order of Nature, 1150–1750*. New York: Zone, 1998.

de Dmitrewski, Michel. "Fr. Bernard Délicieux, O.F.M: Sa lutte contre l'inquisition de Carcassonne et d'Albi, son procès, 1297–1319." *Archivum Franciscanum* 18 (1925).

de Wesselow, Thomas. "The Date of the St. Francis Cycle in the Upper Church of S. Francesco at Assisi: The Evidence of Copies and Consideration of Method." In William Cook, ed., *The Art of the Franciscan Order in Italy*. Leiden: Brill, 2005.

Delaurenti, Beatrice. *La puissance des mots, "Virtus verborum": Débats doctrinaux sur le pouvoir des incantations au Moyen Âge*. Paris: Cerf, 2007.

Delmas, Sophie. "Nicolas de Lyre Franciscain." In Gilbert Dahan, ed., *Nicolas de Lyre: Franciscain du XIVe* siècle, exégète et théologien. Paris: Institut d'Études Augustiniennes, 2011.

DeVun, Leah. *Prophecy, Alchemy, and the End of Time: John of Rupescissa in the Late Middle Ages*. New York: Columbia University Press, 2009.

Diefenbach, Lorenz. *Glossarium Latino-Germanicorum mediae et infimae Aetatis*. Frankfurt: Joseph Baer, 1857.

Diels, H. *Die Entdeckung des Alkohols*. Berlin: Akademie der Wissenschaften, 1913.

Du Cange, Charles Du Frense. *Glossarium mediae et infimae latinatis conditum a Carlo Dufrense Domino du Cange: cum supplementis integris monachorum ordinis S. Benedicti, D.P. Carpenterii et Aditamentis adelungii et aliorum digest G.A.L. Henschel.* Paris: Firmin Didot, 1844.

Eamon, William. "How to Read a Book of Secrets." In Elaine Leong and Alisha Rankin, eds., *Secrets and Knowledge in Medicine and Science, 1500–1800*. Burlington, Vt.: Ashgate, 2011.

———. *Science and the Secrets of Nature*. Princeton, N.J.: Princeton University Press, 1994.

Easton, Stewart. *Roger Bacon and His Search for a Universal Science*. New York: Columbia University Press, 1952; repr. Bridgeport, Conn.: Greenwood, 1970.

Elders, Leo. *Aristotle's Cosmology: A Commentary on the* De Caelo. Assen: VanGorcum, 1965.

Eliade, Mircea. "Alchemy." *Encyclopedia of Religion*, ed. Lindsay Jones. 2nd ed. Detroit: Macmillan, 2005.

———. *The Forge and the Crucible*. 2nd ed. Chicago: University of Chicago Press, 1978.

Elliot, Dyan. "The Physiology of Rapture and Female Spirituality," in Peter Biller and A. J. Minnis, eds., *Medieval Theology and the Natural Body*. Rochester, N.Y.: York Medieval Press, 1997.

Emmerson, Richard. *Antichrist in the Middle Ages*. Seattle: University of Washington Press, 1981.

Fabbri, Christiane Nockels. "Treating Medieval Plague: The Wonderful Virtues of Theriac," *Early Science and Medicine* 12, 3 (2007).

Fairfield, Leslie P. *John Bale, Mythmaker for the English Reformation*. West Lafayette, Ind.: Purdue University Press, 1976.

Firth, Katherine R. *The Apocalyptic Tradition in Reformation Britain*. Oxford: Oxford University Press, 1979.

Forbes, R. J. *A Short History of the Art of Distillation*. Leiden: Brill, 1947.

Fowden, Garth. *The Egyptian Hermes: A Historical Approach to the Late Pagan Mind*. Princeton, N.J.: Princeton University Press, 1986; corrected and with a new preface 1993.

Ganzenmüller, Wilhelm. *Die Alchemie im Mittelalter*. Paderborn: Bonifacius-durckerei, 1938.

Gasquet, F. A. "An Unpublished Fragment of a Work by Roger Bacon." *English Historical Review* 12, 47 (1897).

Getz, Faye. "Roger Bacon and Medicine." In Jeremiah Hackett, ed., *Roger Bacon and the Sciences*. New York: Brill, 1997.

Glorieux, Palémon. *Répertoire des maîtres en théologie de Paris au XIIIe siècle*. Paris: Vrin, 1933.

Gow, Andrew Colin. *The Red Jews: Antisemitism in an Apocalyptic Age*. New York: Brill, 1995.

Grant, Edward. *Planets, Stars, and Orbs.* New York: Cambridge University Press, 1994.

———, ed. *A Source Book in Medieval Science.* Cambridge, Mass.: Harvard University Press, 1974.

Grubmüller, Klaus and Bernhard Schnell, eds. *Vocabularius Ex quo: überlieferungsgeschichtliche Ausgabe.* Tübingen: Niemeyer, 1988–2001.

Hackett, Jeremiah. "Roger Bacon." In Hackett, ed., *Medieval Philosophers*,. Detroit: Gale, 1992.

———. "Roger Bacon: His Life, Career, and Works." In Hackett, ed., *Roger Bacon and the Sciences.* New York: Brill, 1997.

———. "Roger Bacon on *Scientia Experimentalis.*" In Hackett, ed., *Roger Bacon and the Sciences.* New York: Brill, 1997.

Halleux, Robert. *Les Textes alchimiques.* Turnhout: Brepols, 1979.

Halleux, Robert and P. Meyvaert. "Les origines de la *Mappae clavicula.*" *Archives d'Histoire Doctrinale et Littéraire du Moyen Âge* 54 (1987).

Hill, Edmund, O.P. "General Introduction." In Augustine of Hippo, *On Genesis.* Hyde Park, N.Y.: New City Press, 2002.

Hissette, Roland. *Enquête sur les 219 articles condamnés à Paris le 7 mars 1277.* Louvain: Publications Universitaires, 1977.

Holmyard, Eric. *Alchemy.* Harmondsworth: Penguin, 1957.

Hyman, Arthur. *Averroës'* "De substantia orbis." Cambridge, Mass.: Medieval Academy of America, 1986.

Ingham, Patricia Clare. *The Medieval New: Ambivalence in an Age of Innovation.* Philadelphia: University of Pennsylvania Press, 2015.

Jacquart, Danielle. "Le soleil, la lune et les états du corps humain." *Micrologus* 12 (2004).

Johnson, Timothy J. "Introduction: The Franciscan Fascination with the Word." In Johnson, ed., *Franciscans and Preaching.* Boston: Brill, 2012.

Karpenko, Vladimír. "Alchemical View of the Origin of Metals and Transmutation." In Michel Bougard, ed., *Alchemy, Chemistry and Pharmacy*, Proceedings of the XXth International Congress of History of Science (Liège, 20–26 July 1997). Turnhout: Brepols, 2002.

Kieckhefer, Richard. *Magic in the Middle Ages.* New York: Cambridge University Press, 2000.

Kienzle, Beverly Mayne. *Hildegard of Bingen and Her Gospel Homilies.* Turnhout: Brepols, 2009.

Krey, Philip D. W. "The Apocalypse Commentary of 1329." In Krey and Lesley Smith, eds., *Nicholas of Lyra: The Senses of Scripture.* Boston: Brill, 2000.

Lambert, Malcolm. *The Cathars.* Malden, Mass.: Wiley-Blackwell, 1998.

———. *Franciscan Poverty*, rev. and expanded ed. St. Bonaventure, N.Y.: Franciscan Institute, St. Bonaventure University, 1998.

———. Lafleur, Claude with Joanne Carrier, eds. *L'enseignement de la philosophie au XIIe siècle: Autour du "Guide de l'étudiant" du ms. Ripoll 109.* Actes du Colloque International Édités, avec un Complément d'Études et de Textes. Turnhout: Brepols, 1997.

Leggatt, Stuart. "Introduction." In Aristotle, *On the Heavens I & II*. Warminster: Aris & Phillips, 1995.
Leong, Elaine and Alisha Rankin. "Introduction." In Leong and Rankin, eds., *Secrets and Knowledge in Medicine and Science, 1500–1800*. Burlington, Vt.: Ashgate, 2011.
Lerner, Robert E. "Antichrists and Antichrist in Joachim of Fiore." *Speculum* 60, 3 (1985).
———. *The Feast of Saint Abraham: Medieval Millenarians and the Jews*. Philadelphia: University of Pennsylvania Press, 2001.
———. "Historical Introduction." In John of Rupescissa, *Liber secretorum eventuum*, ed. Lerner and Christine Morerod-Fattebert. Fribourg: Éditions Universitaires, 1994.
Little, A. G. *Franciscan Papers, Lists, and Documents*. Manchester: Manchester University Press, 1943.
Lu Gwei-Djen, Joseph Needham, and Dorothy Needham. "The Coming of Ardent Water." *Ambix* 19 (1972).
Maccagnolo, Enzo. "David Dinant and the Beginnings of Aristotelianism in the West." In Peter Dronke, ed., *A History of Twelfth-Century Western Philosophy*. Cambridge: Cambridge University Press, 1988.
Macy, Gary. *Treasures of the Storeroom: Medieval Religion and the Eucharist*. Collegeville, Minn.: Liturgical Press, 1999.
Mandonnet, Pierre. *Siger de Brabant et l'averroïsme latin au XIIIme siècle*. Louvain: Institut Supérieur de Philosophie de l'Université, 1908–1911.
Mandosio, Jean-Marc. "La création verbale dans l'alchimie latine du Moyen Âge." *Bulletin du Cange* 63 (2005): 137.
Marrone, Stephen P. *A History of Science, Magic, and Belief: From Medieval to Early Modern Europe*. New York: Palgrave, 2015.
———. *In the Light of Thy Countenance: Science and Knowledge of God in the Thirteenth Century*. Boston: Brill, 2001.
Martel, Phillipe. *Les cathares et l'histoire: Le drame cathare devant ses historiens, 1820–1992*. Toulouse: Privat, 2002.
Matus, Zachary A. "Alchemy and Christianity in the Middle Ages." *History Compass* 10, 12 (2012).
———. "Reconsidering Roger Bacon's Apocalypticism in Light of his Alchemical and Scientific Thought." *Harvard Theological Review* 2 (2012).
———. "Resurrected Bodies and Roger Bacon's Elixir." *Ambix* 4 (2013).
Mauro, Vincenzo. "La disputata 'de anima' tra Vitale du Four e Pietro di Giovanni Olivi." *Studi Medievali* 38, 1 (1997).
McGinn, Bernard. *Antichrist: Two Thousand Years of the Human Fascination with Evil*. New York: Columbia University Press, 2000.
———. *The Calabrian Abbot: Joachim of Fiore in the History of Western Thought*. New York: Macmillan, 1985.
———. "Pastor Angelicus." *Santi e santità nel secolo XIV: atti del XV Convegno Internazionale, Assisi, 15–16–17 ottobre 1987*. Società Internazionale di Studi Francescani. Convegno Internazionale (15th,1987, Assisi, Italy). Perugia: Università degli studi di Perugia, Centro di Studi Francescani, 1989.

McVaugh, Michael. "The 'Experience-Based Medicine' of the Thirteenth Century." *Early Science and Medicine* 14 (2009).

———."Theriac at Montpellier 1285–1325." *Sudhoffs Archiv, Zeitschrift für Wissenschaftsgeschichte* 56, 2 (1972).

Meyer, Heinz. *Die Enzyklopädie des Bartholomäus Anglicus: Untersuchungen zur Überlieferungs- und Rezeptionsgeschichte von* De Proprietatibus Rerum. München: Fink, 2000.

Migliorino, Francesco. "Alchimia lecita e illecita nel Trecento." *Quaderni medievali* 11 (1981).

Moloney, Brian. *Francis of Assisi and His "Canticle of Brother Sun" Reassessed.* New York: Palgrave Macmillan, 2013.

Moneti, E. G. Muzzioli, I. Rossi, and M. Zamboni, eds. *Catalogo dei Manoscritti della Biblioteca Casanatense.* Roma: Libreria dello Stato, 1949.

Monti, Dominic, O.F.M. "Introduction." In Bonaventure, *Breviloquium*, ed. and trans. Monti. St. Bonaventure, N.Y.: Franciscan Institute, 2005.

Moore, R. I. *The War on Heresy.* Cambridge, Mass.: Harvard University Press, 2012.

Moorman, John. *The Sources for the Life of St. Francis of Assisi.* Manchester: Manchester University Press, 1940.

Moyer, Ian. "Thessalos of Tralles and Cultural Exchange." In Scott Noegel, Joel Walker, and Brannon Wheeler, eds., *Prayer, Magic, and the Stars in the Ancient and Late Antique World.* University Park: Pennsylvania State University Press, 2003.

Multhauf, Robert. "John Rupescissa and the Origin of Medical Chemistry." *Isis* 45 (1954).

Nairn, Thomas A., O.F.M. "St. Francis of Assisi's *Canticle of Creatures* as an Exercise of the Moral Imagination." In Kathleen Hughes, R.S.C.J., ed., *Finding Voice to Give God Praise: Essays in the Many Languages of Liturgy.* Collegeville, Minn.: Liturgical Press, 1998.

Neff, Amy. "Lesser Brothers: Franciscan Mission and Identity at Assisi." *Art Bulletin* 88 (2006).

Nelli, René. *Écritures cathares,* nouv. éd. actualisée et augmentée par Anne Brenon. Monaco: Rocher, 1995.

Newman, William R. "An Overview of Roger Bacon's Alchemy." In Jeremiah Hackett, *Roger Bacon and the Sciences.* New York: Brill, 1997.

———. "The Philosopher's Egg: Theory and Practice in the Alchemy of Roger Bacon." *Micrologus* 3 (1995).

———. *Promethean Ambitions: Alchemy and the Quest to Perfect Nature.* Chicago: University of Chicago Press, 2004.

———. "Technology and Alchemical Debate in the Late Middle Ages." *Isis* 80, 3 (1989).

Nummedal, Tara E. *Alchemy and Authority in the Holy Roman Empire.* Chicago: University of Chicago Press, 2007.

———. "Words and Works in the History of Alchemy." *Isis* 102, 2 (2011).

Obrist, Barbara. "Alchimie et allégorie scripturaire au Moyen Age." In Gilbert Dahan and Richard Goulet, eds., *Allégorie des poètes, allégorie des philosophes: Études sur la poétique et l'hermeneutique de l'allégorie de l'antiquité à la réforme.* Paris: Vrin, 2005.

Odier, Jeanne Bignami. *Études sur Jean de Roquetaillade (Johannes de Rupescissa).* Paris: Vrin, 1952.

Paravicini Bagliani, Agostino. "Ruggero Bacone e l'alchimia della lunga vita." In Chiara Crisciani and Agostino Paravicini Bagliani, eds., *Alchimia e medicina nel Medioevo.* Firenze: Sismel, Edizioni del Galluzzo, 2003.

———. *The Pope's Body.* Trans. David S. Peterson. Chicago: University of Chicago Press, 2000.

Partington, J. R. "Albertus Magnus on Alchemy." In Allen G. Debus, *Alchemy and Early Modern Chemistry: Papers from Ambix.* [London]: Jeremy Mills, 2004.

Patton, Corrine. "'Creation, Fall and Salvation: Lyra's Commentary on Genesis 1–3." In Philip D. W. Krey and Lesley Smith, eds., *Nicholas of Lyra: The Senses of Scripture.* Boston: Brill, 2000.

Pereira, Michela. *The Alchemical Corpus Attributed to Raimond Lull.* London: Warburg Institute, 1989.

———."Alchemy and the Use of Vernacular Languages in the Late Middle Ages." *Speculum* 74, 2 (1999).

———. "L'Alchimista come medico perfetto nel *Testamento* pseudolulliano." In Chiara Crisciani and Agostino Paravicini Bagliani, *Alchimia e medicina nel Medioevo.* Firenze: SISMEL Editzioni del Galluzzo, 2003.

———. "Cosmologie alchemiche." In *Cosmogonie e cosmologie nel medioevo: Atti del Convegno della Società Italiana per lo Studio del Pensiero Medievale (S.I.S.P.M.), Catania, 22–24 settembre 2006*, cur. Concetto Martello, Chiara Militello, and Andrea Vella. Louvain-la-Neuve: FIDEM, 2008.

———. "I francescani e l'alchimia." In Paolo Capitanucci, ed., *I Francescani e le scienze.* Spoleto: Centro Italiano di Studi sull'Alto Medioevo, 2008.

———. "Heavens on Earth: From the Tabula Smaragdina to the Alchemical Fifth Essence." *Early Science and Medicine* 5, 2 (2000).

Peters, Edward. *The Magician, the Witch, and the Law.* Philadelphia: University of Pennsylvania Press, 1978.

Power, Amanda. "A Mirror for Every Age: The Reputation of Roger Bacon." *English Historical Review* 121, 492 (June 2006).

Principe, Lawrence M. "Alchemy Restored." *Isis* 102, 2 (2011): 305–12.

———. *The Secrets of Alchemy.* Chicago: University of Chicago Press, 2013.

Principe, Lawrence M. and William R. Newman. "Some Problems with the Historiography of Alchemy." In William R. Newman and Anthony Grafton, eds., *Secrets of Nature: Astrology and Alchemy in Early Modern Europe.* Cambridge, Mass.: MIT Press, 2001.

Reeves, Marjorie. *The Influence of Prophecy in the Later Middle Ages: A Study in Joachism.* Notre Dame, Ind.: University of Notre Dame Press, 1993.

———. *Joachim of Fiore and the Prophetic Future: A Medieval Study in Historical Thinking,* rev. ed. Stroud: Sutton, 1999.

Rouse, Richard H. and Mary A. Rouse. *Preaching, Florilegia, and Sermons: Studies on the* Manipulus Florum *of Thomas of Ireland.* Toronto: PIMS, 1979.

Sarton, George. *Introduction to the History of Science.* Baltimore: Williams and Wilkins, 1927–48.

Seligman, Adam, Robert Weller, Michael Puett, and Bennett Simon. *Ritual and Its Consequences.* New York: Oxford University Press, 2008.

Şenocak, Neslihan. *The Poor and the Perfect: The Rise of Learning in the Franciscan Order, 1209–1310.* Ithaca, N.Y.: Cornell University Press, 2012.

Shogimen, Takashi. "Academic Controversies." In G. R. Evans, ed., *The Medieval Theologians.* Oxford: Blackwell, 2001.

Siraisi, Nancy G. *Medieval and Early Renaissance Medicine: An Introduction to Knowledge and Practice.* Chicago: University of Chicago Press, 1990.

Smith, Cyril Eugene. *The University of Toulouse in the Middle Ages: Its Origins and Growth to 1500 A.D.* Milwaukee: Marquette University Press, 1958.

Smith, Jonathan Z. "The Bare Facts of Ritual." In Smith, *Imagining Religion: From Babylon to Jonestown.* Chicago: University of Chicago Press, 1982.

———. "The Temple and the Magician." In Smith, *Map Is Not Territory.* Leiden: Brill, 1978.

Smith, Pamela. "Secrets and Craft Knowledge in Early Modern Europe." In Elaine Leong and Alisha Rankin, eds., *Secrets and Knowledge in Medicine and Science, 1500–1800.* Burlington, Vt.: Ashgate, 2011.

Solignac, Laure. *La théologie symbolique de saint Bonaventure.* Paris: Parole et Silence, 2010.

Sorrell, Roger D. *St. Francis of Assisi and Nature: Tradition and Innovation in Western Christian Attitudes Toward the Environment.* New York: Oxford University Press, 1988.

Theissen, Wilfrid. "The Attraction of Alchemy for Monks and Friars in the 13th–14th Centuries." *American Benedictine Review* 46, 3 (1995).

Thijssen, J. M. M. H. *Censure and Heresy at the University of Paris, 1200–1400.* Philadelphia: University of Pennsylvania Press, 1998.

Thompson, Augustine, O.P. *Francis of Assisi: A New Biography.* Ithaca, N.Y.: Cornell University Press, 2012.

Thorndike, Lynn. *History of Magic and Experimental Science.* New York: Columbia University Press, 1923–1958.

Thorp, John. "The Luminousness of the Quintessence." *Phoenix* 36, 2 (1982).

van Bladel, Kevin. *The Arabic Hermes: From Pagan Sage to Prophet of Science.* New York: Oxford University Press, 2009.

Van Engen, John. *Rupert of Deutz.* Berkeley: University of California Press, 1983.

Van Steenberghen, Fernard. *Aristotle in the West: The Origins of Latin Aristotelianism.* 2nd ed. Trans. Leonard Johnston. Louvain: Nauwelaerts, 1970.

Voigts, Linda Ehrsam. "Plants and Planets: Linking the Vegetable with the Celestial in Late Medieval Texts." In Peter Dendle and Alain Touwaide, eds., *Health and Healing from the Medieval Garden.* Woodbridge: Boydell, 2008.

Whalen, Brett. *Dominion of God: Christendom and the Apocalypse in the Middle Ages.* Cambridge, Mass.: Harvard University Press, 2009.

Williams, Steven J. "Roger Bacon and the *Secret of Secrets*." In Jeremiah Hackett, ed., *Roger Bacon and the Sciences*. New York: Brill, 1997.

———. *The* Secret of Secrets*: The Scholarly Career of a Pseudo-Aristotelian Text in the Latin Middle Ages*. Ann Arbor: University of Michigan Press, 2003.

Wilson, C. Anne. *Philosophers, Iōsis and Water of Life*. Leeds: Leeds Philosophical and Literary Society, 1984.

Wippel. John F. "The Condemnations of 1270 and 1277 at Paris." *Journal of Medieval and Renaissance Studies* 7, 2 (1977).

———. "The Parisian Condemnations of 1270 and 1277." In Jorge J. E. Gracia and Timothy B. Noone, eds., *A Companion to Philosophy in the Middle Ages*. Oxford: Blackwell, 2003.

Woodward, David with Herbert M. Howe. "Roger Bacon on Geography and Cartography." In Jeremiah Hackett, ed., *Roger Bacon and the Sciences*. New York: Brill, 1997.

Zanier, Giancarlo . "Procedimenti farmacologici e pratiche chemioterpeutiche nel *De consideratione quintae essentiae*." In Chiara Crisciani and Agostino Paravicini Bagliani, eds., *Alchimia e medicina nel Medioevo*. Firenze: SISMEL Editzioni del Galluzzo, 2003.

INDEX

ACKNOWLEDGMENTS

There are many people and organizations whose support was crucial to the development and completion of this book. Through the first phase of this project, Kevin Madigan, Beverly Kienzle, and Anne Monius were critical cheerleaders. A Fulbright-Hays award enabled research in Italy. Staffs at the libraries of Harvard University, the American Academy at Rome, and Boston College were very helpful. I am also grateful to the librarians at various rare book libraries including the Houghton Library at Harvard, the Biblioteca Casanatense and Biblioteca Angelica in Rome, and the Beinecke Library at Yale. Andrea Frank and Nina Bogdanovsky at Boston College deserve mention for their assistance with images. I also owe a special thank you to two anonymous readers of my manuscript. They provided me a critical, yet collegial shove when I needed one. Likewise I am grateful to Jerry Singerman at the University of Pennsylvania Press for his patience and helpful critique. I also would like to thank my colleagues at Boston College for their support and their belief in my ability to bring this project to a fruitful conclusion. I would also like to thank the students who have taken one or another iteration of my "Science before the Scientific Revolution" course. The questions and assumptions of these bright young undergraduates have helped me reexamine my own. Finally, to my wife and my children, I owe a special thank you, and I am hopeful that this book will serve my boys as a piece of material evidence that I do something of meaning when I am not with them even if they can write several books complete with illustrations per week.

I lay exclusive claim, however, to any and all errors that reside within the text. And, while my biases are likely least visible to me, I have tried to make my own assumptions and convictions as plain as possible.